Thriving as an International Scientist

The publisher and the University of California Press Foundation gratefully acknowledge the generous support of the Atkinson Family Foundation Imprint in Higher Education.

Thriving as an International Scientist

Professional Development for Global STEM Citizens

Sonali Majumdar, PhD

UNIVERSITY OF CALIFORNIA PRESS

University of California Press
Oakland, California

Library of Congress Cataloging-in-Publication Data

Names: Majumdar, Sonali, author.
Title: Thriving as an international scientist : professional
    development for global STEM citizens / Sonali
    Majumdar.
Description: Oakland, California : University of
    California Press, [2025] | Includes bibliographical
    references and index.
Identifiers: LCCN 2025011328 (print) | LCCN 2025011329
    (ebook) | ISBN 9780520401037 (cloth) | ISBN
    9780520401044 (paperback) | ISBN 9780520401051
    (ebook)
Subjects: LCSH: Science—Vocational guidance—United
    States—21st century. | Technology—Vocational
    guidance—United States—21st century. |
    Engineering—Vocational guidance—United
    States—21st century. | Mathematics—Vocational
    guidance—United States—21st century. | Graduate
    students, Foreign—Vocational guidance—United
    States—21st century.
Classification: LCC Q149.U5 M315 2025 (print) | LCC Q149.U5
    (ebook) | DDC 702.3/73—dc23/eng/20250421
LC record available at https://lccn.loc.gov/2025011328
LC ebook record available at https://lccn.loc.gov
    /2025011329

GPSR Authorized Representative: Easy Access System
Europe, Mustamäe tee 50, 10621 Tallinn, Estonia, gpsr.
requests@easproject.com

34 33 32 31 30 29 28 27 26 25
10 9 8 7 6 5 4 3 2 1

To my parents, who gave me confidence to explore and imagine.
Now I imagine what they would say about the book.

CONTENTS

ILLUSTRATIONS

ACKNOWLEDGMENTS

I am sincerely thankful to many individuals for their guidance, support, contributions, and feedback that shaped this book.

I owe immense gratitude to Sina Safayi, without whom this book would not have been written. After advocating for international graduate students and postdocs for many years, Sina discussed the need for a specialized career development guidebook for international scientists with the editorial staff at UC Press. Sina trusted me to implement this vision and connected me with the press. Moreover, he gave me the idea to write this book when I needed it the most, a few months after suffering a great personal loss. Trying something completely new was exactly the challenge I needed at the time. I hope I met that challenge.

This book is for international scientists, by international scientists. I offer my heartfelt gratitude to the many scientists who heeded my open call to share their insights, experiences, and advice. Beyond exemplifying the concepts outlined, their stories—of determination, courage, and vulnerability—are the real essence of this book. Special thanks to Adriana Bankston, who shared her expertise in science policy and coauthored chapter 13 on systemic reforms to improve scientific training and retain STEM talent in the United States.

While I embarked on this inaugural book-writing adventure with reckless abandon, it was the support of my professional community that gave me strength along the journey. Many thanks to colleagues who shared advice and experiences about the book publishing process. Thank you to peer reviewers, Amy Aines, Chris Smith, Derek Attig, and Natalie Chernets, whose thoughtful comments significantly improved the quality of the proposal and the manuscript. I also thank the editorial committee of UC Press for their generous support and encouragement, which further validated the importance of this topic.

I couldn't be more thankful for the patience, guidance, and consistent support of UC Press staff, particularly Chloe Layman, my acquisition editor whose affirmation and guidance was critical throughout my writing. I would also like to express appreciation to UC Press's First Generation program for uncovering the hidden curriculum of book publishing and walking first-time authors like me through the process in a timely manner.

This book is an ode to, and celebration of, people and communities. To my friends, family, mentors, coworkers, and professional community (particularly those in the Graduate Career Consortium) who cheered for me and encouraged me throughout this process, I am truly humbled. Thank you!

Cheers to coffee, Pink Floyd, and Queen, my faithful companions over the past two years.

PREFACE

I am an international scientist striving to improve PhD and postdoc training in the United States. My and my peers' lived experiences motivated my transition from scientific research to improving training experiences for budding scientists through access to professional development that did not exist during our time. Career and professional development support for early-career scientists in universities has surged in the past decade due to the initiatives and recommendations of national scientific agencies such as the National Science Foundation (NSF), the National Academy of Sciences (NAS), the National Institutes of Health (NIH), and the American Association for the Advancement of Science (AAAS), among others. Many higher education institutions are taking steps to improve training and better support an increasingly diverse population of graduate students and postdocs.

Yet there is a notable dearth of specific advocacy, guidance, and support for international scientists (foreign born, temporary visa holding) in the United States at all career stages. The challenges they face deserve special attention as they make up about 41% of PhD students and about 58% of postdocs in science and engineering (S&E) (NCSES 2025). We internationals also face additional barriers to professional well-being and success due to many factors, including a convoluted and dated immigration system, the hidden curriculum of higher education, the lack of socioeconomic support, and disparities for non-native English speakers and those from the Global South. Another drawback is the lack of representation of international professionals in academic administration, likely as a result of structural (read: immigration) barriers. For example, international professionals are vastly underrepresented in the field of graduate- and postdoc-level career and professional development (CPD). I clawed my way into CPD and broadly academic administration—a field in which typically less than 2% of staff is international. Now, as an educator and administrator, I continue to bring attention to needed support for international scientists while helping other international professionals enter graduate-level CPD. Toward that goal, I mentor international trainees to enter this field and cofounded the International Community of Practice for international professionals within the Graduate Career Consortium (GCC), an organization providing a leading voice in graduate and postdoc career and professional development.

There has been growing acknowledgment of the diverse careers of scientists: research and teaching, business of science, science communication, publishing, science policy, and entrepreneurship, among others. Given the immigration-related challenges, international scientists typically move into research careers in academia or industry and are underrepresented in many other science-related careers. This book provides actionable guidance to international scientists for professional success and well-being during doctoral education and beyond. I discuss lifelong learning, a growth mindset, and creative approaches to success amid systemic barriers; strategic planning and communicating technical and foundational skills; the unique strengths that international scientists' lived experiences render; and the power of the international community for collective learning and empowerment. I also give tactical advice on uncovering the hidden curriculum in professional advancement, conducting job searches in the United States, and parallel planning for global careers. While this book caters primarily to international scientists, section 6, on how to support international employees and trainees and enable systemic changes, is addressed to faculty advisers, career educators, managers, and advocates.

Like many international scientists, I navigated challenging circumstances and personal sacrifices because of the restrictive and xenophobic immigration climate. I weave into the bigger data-informed picture my lived experience and anecdotes from international scientists who ventured into a variety of careers, in and beyond research, to exemplify concepts outlined in each chapter. Specifically, the chapters include examples of actionable ways in which international scientists have applied the highlighted concepts and approaches to navigate barriers and achieve their goals, reimagine professional possibilities, and expand definitions of career success.

This book centers and amplifies the voices and rich experiences of international scientists to best serve fellow international scientists. In academic and other professional circles, international scientists are mostly invoked as either a model minority in the workforce and statistics on research productivity or a monolithic community when citing challenges. In chapters 10 and 11, I highlight the humanity of international scientists, going beyond those narratives to illuminate the sacrifices and human cost of surviving the current immigration system, maintaining legal status, and succeeding in unsupportive systems. But that is only one aspect of our multifaceted lives. This book celebrates international scientists—our strengths, determination, remarkable resilience, and resolute optimism—as we brave unfavorable environments to bring talent and humanity to science. We international scientists represent the global diversity of ethnicity, race, country of origin, socioeconomic status, languages, and, most of all, lived experiences. Our worldview, ingenuity, diverse perspectives, and leadership styles enrich the scientific landscape in academia, industry, and other sectors. Through good and bad times, international scientists rely on community to share knowledge, empower, mentor, and solve problems. Through stories of these international communities, I highlight the power of collective action and communal championship in a predominantly individualistic US society. This book is in equal parts a celebration of the international scientific community—of our adventures and brave journeys—an invitation to fearlessly chart creative futures equipped with tools and strategies, and a call to create an inclusive and equitable society for global citizens of science.

The Case for Tailored Support for International Scientists

Introduction

My Story

My journey as an international scientist began in 2007. I moved from India, where I was born and raised, to enroll in the PhD program at the University of Georgia's Department of Biochemistry and Molecular Biology. I wanted to study the dynamic biological processes driven by RNAs, or ribonucleic acids, after learning about retroviruses (e.g. HIV, coronavirus) in my undergraduate molecular biology course. My PhD experience was like a roller-coaster ride. There were highs in the beginning, but it was during the middle years when I experienced low points. However, I did not entertain thoughts of quitting even at the lowest times or when I was confronted with doubt or pity from others. At the time, I did not know this attribute, resilience, was noteworthy. I simply knew giving up was not an option, a hard-earned life lesson from growing up in hypercompetitive environments in India.

In retrospect, I am certain that my PhD experience would have felt more challenging without my previous work experience. This experience gave me confidence and a healthy perspective while I navigated PhD training. I had worked in healthcare for about two and a half years between finishing my MSc degree in India and starting the PhD program in the United States. I was one of a four-member team of physicians and embryologists who launched an in vitro fertilization (IVF) center in a hospital in India. Due to the start-up environment, I was privy to the entire life cycle of a typical IVF process. In addition to the science, I was exposed to the psychological stress couples face, placing heavy responsibilities on the healthcare team. I also learned about business models in healthcare. Ultimately, the rare opportunity to closely observe and contribute to the conception of life gave me a healthy perspective on what matters. My first job at the IVF clinic remains my most technically and psychologically challenging professional experience, so I knew exhaustion and burnout firsthand. As a result, I maintained healthy work-life boundaries and rarely worked late evenings or weekends during my PhD training.

I approached my PhD with an open-ended career in mind. My post-PhD plans involved conducting research in a biotech or biomedical research institute. While I enjoyed the creativity of research, I did not enjoy grant writing, and I did not want the stress and overwork involved in being an academic faculty researcher. In line with my original plan, I took up a postdoctoral research scholar position at the Sloan Kettering Institute. I selected this position because my PhD research was focused on discovery-based science, and I needed skills and experience in translational science. Postdoc training in New York City would also provide access to biotech professional networks and exposure to careers in biotech. However, a combination of my lived experience, role models, peers and mentors, and the climate and national discourse on scientific training at the time changed my professional interest and future. My volunteer experiences with the Memorial Sloan Kettering Cancer Center's Postdoc Association (MSK-PDA) and the founding board of the New York City Postdoc Coalition (NYCPC) exposed me to the shared challenges of fellow postdocs, primarily lack of creativity in research, inequitable access to professional development and mentorship, low wages for high-skilled labor, psychologically unsafe (and some downright abusive) work environments, lack of financial stability, and a high cost of living. These are symptoms of the systemic and cultural issues in the academic research enterprise, so I became increasingly involved in grassroots volunteer work to improve graduate and postdoc training. At the same time, I attended career development workshops, where I met my role models, Thomas (Tom) Magaldi and Ushma Neil, whose mentorship I actively sought. Tom and Ushma were biomedical scientists who applied their scientific acumen to improve the training of early-career scientists through career and professional development programs. Tom introduced me to a large network of professionals in the field who were members of the Graduate Career Consortium (GCC) and the National Postdoctoral Association (NPA). The professional communities at GCC and NPA gave me a newfound passion and interest in building holistic PhD and postdoc training to empower socially engaged, dynamic, and creative academics.

However, the biggest hurdle to my career transition was the immigration system in the United States. While there are training and work visas for international scientists to pursue research and teaching roles related to their PhDs, visa policies are ambiguous with regard to making career transitions to adjacent fields such as academic administration. Moreover, I was making this career change during the first Trump administration, when the immigration climate was particularly restrictive and employers were largely risk averse. My primary strategy was to focus on what I could control: play the numbers game and impress prospective employers in interviews. Consequently, I stacked up job offers—I had four offers at one point—and negotiated until at least one employer, the University of Virginia (UVA), agreed to sponsor my work visa.

The silver lining of searching for a job as an international is that you are able to gauge the mindset of your prospective colleagues and the broader culture of the organization by the way they handle visa sponsorship. Those who don't shy away from taking risks on sponsorship also appreciate bold and innovative work. As my work involves building new avenues for systemic improvements, I am more likely to succeed in places that encourage innovative ideas and embrace risks. UVA's Office of Graduate and Postdoctoral Affairs was that workplace for me. We made a compelling case for the relevance of my direct knowledge of graduate and postdoctoral training and scientific skills, and my visa petition was accepted without issue.

My role at UVA involved working again in a start-up environment at a historic institution. To actualize a bold vision afforded by institutional investment, I assumed responsibility for developing and implementing a new professional development program called PhD Plus, intended to secure long-term career success for *all* PhD students and postdocs at UVA. But my heart and passion lay with early-career scientists. PhD (and postdoc) training in STEM has unique systemic issues and challenges, and I wanted to devote all my energy to improving scientific training. This led me to my current role as assistant dean for professional development in the Graduate School at Princeton University, where I strive for equitable access to professional development among graduate students through the GradFUTURES program, with a special focus on the natural sciences and engineering disciplines.

By now, it should be obvious that my career has been nonlinear. But every professional experience has influenced my path and development. I have discovered as much about myself as I have about science, people, and societies. The PhD experience has changed me in ways that I continue to discover. It trains us to think, learn, discover, and persist. Among other skills, I use a superpower from my scientific training in my current work: pattern recognition. I recognized patterns in data for scientific explorations before; now I apply it to developing strategies, building networks and consensus, and assessing programs.

Even as I firmly believe in the broad value and utility of scientific training, I grieved about no longer discussing, dabbling in, and learning about RNAs. The COVID-19 pandemic came with a fresh surge of grief. I had pursued RNA research as I was intrigued by viruses, but I was not in the trenches with my fellow scientists studying the coronavirus or mRNA vaccines. The veering away from "topic land" can come with a loss of identity (Mullaney and Rea 2022): Can I still identify as a scientist if I am not actively doing bench science? In fact, I still use my science skills and identify as a scientist. "Scientist" is who we are, not what we do. It is the way of being and applying the scientific method to diverse problems and fields that makes us scientists.

It is not easy living with our evolving motivations, interests, and persons. Following our interests and passions comes at huge personal costs to us and our families. We miss family milestones regularly—weddings, births, sometimes deaths. Our loved ones are forced to reckon with our absence while we are fixed on scientific problems continents away. We rely on ourselves and our friends to navigate most challenges without proper support systems. I navigated surgeries with the help of a few friends while my family in India worried from afar.

Immigration rules and policies make most things hard—from following diverse interests and careers to taking extended breaks or even traveling home regularly. In the past seventeen years, I took no extended breaks—neither before my PhD or postdoc nor between my jobs at UVA and Princeton—due to the uncertainty of visas. Securing work permits and visas can sometimes take a while given various hurdles and background checks. So, because of the need to plan early and strategically to navigate this maze, planning a hiatus (which is unpaid) is either economically infeasible or not conducive to maintaining your immigration status. And if you are from China or India, the state of misery can be prolonged for decades. Like many Indians, my immigration journey was long, convoluted, and complex. I continued to make incremental progress toward permanent residency. UVA filed a green card petition on my behalf in 2020 under Employment Based category 2 (EB-2). While my petition was approved, I couldn't get a green card because I am an Indian citizen. The wait time under EB-2 for Indian citizens is about ten years. However, I had secured my place in line (called a priority date). I could get my green card faster if I upgraded to the EB-1 category. So I kept building my portfolio to be competitive for EB-1A. In 2024, Princeton University filed an EB-1A petition on my behalf (importing my 2020 priority date). The legal team carefully developed a compelling case over six months. My petition was approved in six days, and I got my green card six months later. I finally became a permanent resident in 2024, after living in the United States for seventeen years.

The xenophobic climate of recent years, with travel bans and backlogs, triggered by politicians and the pandemic, has added to the trauma and tested the endurance of immigrants. I once went to a memorial for an Iranian student and was equally shocked by their circumstances and touched by the solidarity of the Iranian community. As Iranian immigrants in the United States have endured travel bans for almost a decade, some haven't gone home for many years. They rely on each other through successes, tragedies, and everything in between. Through the COVID-19 pandemic and climate disasters, we watched helplessly while our families and communities at home suffered. Or we rallied to provide support and resources from afar. I lost my parents in 2022; my mother passed away in March after battling chronic illness, and my father passed away suddenly in August. I was stuck in the United States due to visa backlogs and couldn't join the rest of my family for their last rites. It was the lowest point of my life, and I continue to process grief, guilt, and other emotions. Unfortunately, I am not an exception. Many immigrants share similar misfortunes. Consequently, beneath the surface of our work ethic, we carry burdens and complicated feelings for daring to embark on this journey. We tend to carry the battle scars silently and rely on our superpowers, endurance and community.

The second Trump administration will bring more chaos and hardships for immigrants. We will continue to cope, relying on muscle memory and community solidarity. In a fairer world, internationals would not face such systemic barriers and discrimination because we enrich our environments while overcoming steep odds. We deserve dignity and respect.

CHAPTER ONE

The Critical Need for Specialized Career and Professional Development of International Scientists

International scientists need specialized career and professional development support that considers our diverse demographic backgrounds, the lack of established social and professional support networks, the hidden curriculum of life and career in the United States, specific circumstances, the barriers and challenges such as inadequate access to independent research funding streams, the unfavorable sociopolitical climate and immigration system, travel restrictions, and the vulnerability to workplace discrimination. The US immigration system presents a potent barrier to the career agency of international scientists. International PhDs either fight uphill battles to enter fields that are not closely related to their graduate or postdoc research topics or plan their careers around immigration feasibility. We seek ad hoc support, often from international peer communities or well-intentioned but helpless and overworked educators and career advisers. In this chapter, I clarify the issues faced by international scientists to contextualize concepts and approaches outlined in subsequent chapters. I use a data-informed lens to indicate the growth of international scientists in STEM fields and highlight the importance of international talent in maintaining US dominance in scientific research. I also outline sources of disparity in scientific training and inequities that influence career decisions, which ultimately has an impact on the scientific enterprise.

DATA ON INTERNATIONAL SCIENCE TRAINEES AND DOCTORAL RECIPIENTS OVER THE YEARS

US universities have seen significant growth in their international graduate student and postdoc populations (temporary visa holders). According to the NSF Survey of Graduate Students and Postdoctorates in science, engineering and health (GSS, table 1–3a), in 2023, internationals (temporary visa holders) made up 40.7% (n = 125,030) of PhD students and 57.9% (n = 38,149) of postdocs in the United States. Moreover, between 1980 and 2023, the international graduate student population (master's and PhDs) grew from 13.7% to 39.4%, amounting to an 25.7% increase (NCSES 2025).

International enrollments at American colleges rebounded strongly during the 2022–23 academic year, reflecting a 21% increase in international graduate student enrollment from the previous year (Fischer 2023). This increase brings the number of international graduate students enrolled in master's and PhD degree programs in 2023 across all fields in the United States to 467,000, an all-time high. Similarly, between 1980 and 2023 the international postdoc population grew from 35.4% to 57.9%, a 22.5% increase (NCSES 2025). Academic scientific research has benefited tremendously from the contributions of international trainees. At the same time, US universities maintain global leadership in recruiting and training scientists.

International scientists trained in the United States are also key to maintaining the country's leadership and innovation in S&E fields. NSF's 2022 Survey of Earned Doctorates (SED) indicates a significant growth in temporary visa holders earning a PhD degree (NCSES 2023, table 1-6). International scientists made up 11% of PhD graduates in 1977 and 34% in 2022. Even considering the proportionate increase in the overall number of PhD recipients between 1977 and 2022, international

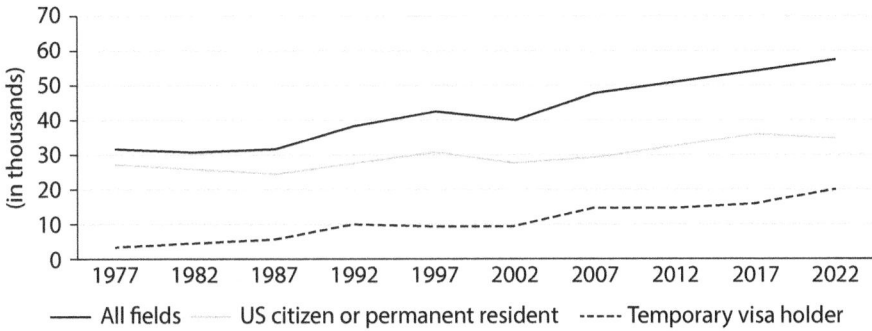

Figure 1. Doctoral recipients by citizenship status, 1977–2022 (NCSES 2023, table 1-6).

scientists account for most of the increase (62%, or 16,144 of 25,880). Figure 1 indicates the similarity in trend lines between the overall increase in PhDs in all fields and temporary visa holders, particularly since 2002.

The majority of international S&E PhD recipients in the United States were born in China and India (NCSES 2023, table 7-7). This is important to highlight as these scientists face additional challenges, such as a longer time to achieve permanent residency status. There is a high opportunity cost of inadequate career development and lack of immigration support for international scientists, particularly those facing overwhelming challenges. The United States risks losing scientific talent that it trains, to a competitive global market.

SYSTEMIC BARRIERS TO THE TRAINING AND CAREER DECISIONS OF INTERNATIONAL SCIENTISTS
Sources of Funding among International Scientists

The source of funding influences training experience, independence, and flexibility and ultimately the career decisions of PhD students and postdocs in S&E. Among international PhD students in S&E, 51.8% are funded primarily through research assistantships or traineeships

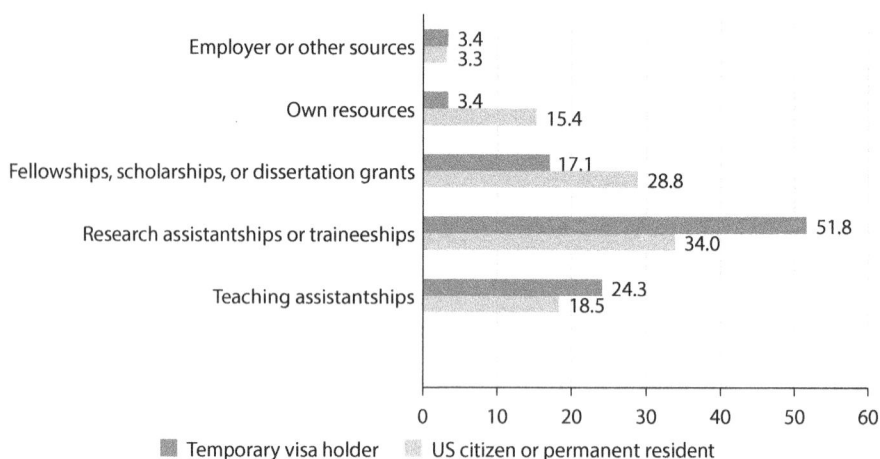

Figure 2. Primary source of funding for doctoral recipients in science and engineering (in percent) by citizenship (NCSES 2023, table 4-1), temporary visa holder (n = 15,935) and US citizen or permanent resident (n = 24,917).

(NCSES 2023, table 4-1) (figure 2). In comparison, 34% of S&E PhD students who are US citizens or permanent residents are funded through research assistantships and 28.8% receive funding through fellowships, scholarships, and dissertation grants. As research assistantships are typically funded under a faculty or principal investigator's research grant, for the most part, international PhD students rely on their faculty advisers for funding. The reliance on funding restricts the ability of international scientists to explore independent and creative research ideas and makes them vulnerable to harassment and bullying in research labs (Fleming 2022).

The data on primary sources of funding among PhD recipients from US universities who continue postdoc training tell a similar story (NCSES 2023, table 2-5): 53.1% of international PhD graduates of US universities in 2022 relied on institutional funding for postdoc training. compared to 36.2% of US citizens and permanent residents. Only 18% of international PhDs, as compared to 37.5% of US citizens and permanent residents, received government funding such as fellowships.

Collectively, these data indicate that fewer international graduate students and postdocs receive independent fellowships. A key reason is lack of access; international PhD students and postdocs in S&E are eligible for only a small handful of independent fellowships, primarily from nonfederal sources. This has substantial disadvantages regarding competitiveness for tenure-track faculty positions, where grantsmanship success is typically a proxy for competitive independent research ideas. Therefore, international scientists need guidance on creative ways to access diverse research funding opportunities. In chapter 3, I introduce a framework for creative thinking and exploration citing research funding as a test case. While creative thinking can certainly help navigate restrictive environments in the short term, the burden falls on international trainees. Ultimately, international PhD students and postdocs warrant specific considerations in systemic enhancements in graduate education and postdoc training (see chapter 13).

Career Decisions of International Scientists

The 2023 NSF Survey of Earned Doctorates compares the percentage of recent international PhD graduates who committed to either continuing postdoc training or securing employment in academia, industry, government, or the nonprofit sector (except postdocs) over the past thirty years (NCSES 2023, table 2-7) (figure 3). Overall, among the temporary visa–holding PhD population across the years, a higher percentage (~60.5% in 2022) seek employment rather than continued postdoc training. When comparing the proportionate population, international PhDs in the workforce were about 2.5 times higher in 2022 than 1992 (59.3% of 13,686, or 8,115, in 2022 vs. 60.5% of 5,480, or 3,315, in 1992).

Since 1997, the majority of international PhD graduates have been employed in industry or business (i.e., for-profit or corporate sectors). In 2022, 72.3% of international PhD graduates who were not doing postdoc research were hired in industry, most likely in R&D fields. An interesting observation of the first-destination outcome is that in 1992,

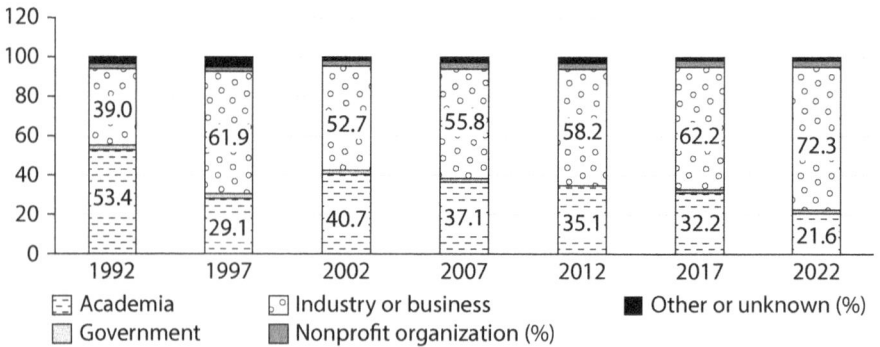

Figure 3. Percent of international (temporary visa-holder) doctoral recipients reporting employment commitments by sector, 1997–2022 (NCSES 2023, table 2-7).

53.4% of the international PhDs were employed in academia (most likely research or teaching-related faculty positions), compared to 21.6% in 2022.

While academic faculty positions decreased, there was an increase in the number of industry R&D positions, which offer better wages and benefits. This led to declining interest in continuing postdoc training among PhDs, irrespective of citizenship. More importantly, for international PhDs graduates of US universities, the STEM OPT (Optional Practical Training) extension policy was passed in 2008 (Thompson 2012). The OPT is a means of temporary employment for international students under the US F-1 (student) visa status. The STEM OPT extension that passed in 2008 provided three years (compared to one year previously) of postgraduate employment options in diverse sectors to international PhD graduates in STEM fields. It also gave employers more time to sponsor work visas (typically called H-1B or O-1 visas) for international scientists. Even so, the percent change in industry outcomes among international PhD grads since 2017 can be attributed in part to the tech boom triggered by the pandemic, which is already showing a downturn and will likely experience further slowdown in coming years.

There is still a lack of real career diversity among international PhDs. Less than 3% of international PhDs are hired in the nonprofit sector and less than 2% in the government sector. It is likely that the nonprofit sector cannot compete for international PhD talent due to the lack of infrastructure (financial, HR, and immigration) required to support them. Consequently, the US STEM enterprise does not fully utilize the capabilities of international talent or lacks global diversity of perspectives, particularly in fields of significant import such as science policy.

The visa stipulations also make it challenging for international PhDs to pursue roles that are not closely related to their doctoral research topics. I speak from personal experience. I had to beat steep odds to transition from research to higher education administration. The immigration system, until recently (chapter 7, H-1B Modernization Rule), has also disincentivized entrepreneurship among international scientists. The burden of finding creative ways to access non-research careers still lies solely with the individual international scientist. This is particularly true of citizens of China and India. Chinese and Indian scientists face backlogs and long wait times to get permanent residency status. Moreover, the backlash against immigration and rising discrimination against Asians pose additional barriers to recruiting, retaining, and growing that talent pool. Faced with a hostile climate and lacking diverse opportunities in the United States, increasing numbers of international scientists from China (Xie et al. 2023) and India are exploring other countries for PhD training as well as opportunities for global mobility after earning their PhDs in the United States. Therefore, lack of access to diverse career opportunities for international scientists challenges the United States's standing as a global force for scientific innovation. To this end, the federal government must take corrective actions. In particular, immigration-related systemic changes are essential for long-term sustainable career diversity among international scientists in the United States.

International scientists fall prey to deficit thinking because of regularly operating within restrictive environments and systems. We must actively fight our circumstances to shape our ambitions and expand our possibilities. Even so, the volatile labor market can pose challenges and threats to the already precarious circumstances that international scientists navigate. Job declines in specific areas can disproportionately impact international scientists who trained in those fields and may not be able to pivot easily due to visa stipulations. International scientists also face special risks during large-scale layoffs, as seen recently in the tech sector. As visa rules stipulate, international scientists have a short grace period (typically two months) to find employment or face deportation.

Gertrude Nonterah faced such circumstances and relied on her resourcefulness and creativity to actively reinvent herself. She came to the United States from Ghana to study biology on a pre-med track, but she soon realized that pursuing medical school was untenable for international students, as it was expensive and international students typically have to pay for it themselves or find a cosigner for loans. Consequently, she used a clever strategy to explore a parallel track, an accelerated nursing program. It was similar to pursuing a career as a doctor, and because of a shortage of healthcare workers, international students had a viable path to permanent residency. She thoroughly enjoyed her experience in the accelerated nursing program. But to her dismay, labor market trends changed. By the time she graduated, the need for trained nurses had been met. As a result, it became difficult to secure full-time nursing jobs, so she found herself at another crossroad. While exploring nursing, she had contemplated pursuing a PhD. She revisited her original intent and pursued a PhD and then postdoc training. Little did she know that the majority of STEM PhDs do not succeed in getting a tenure-track faculty position. Her postdoc ended abruptly when her principal investigator ran out of funding (another risk that international scientists disproportionately bear due to fewer independent fellowships). She applied for many faculty positions and second postdocs but got no responses. At least, during her postdoc, she was able to secure permanent residency status.

Gertrude then seriously started exploring nonacademic careers. During her postdoc, she did freelance writing for a year. She wrote grants (beyond science), articles, and business and financial blog posts. In addition to expanding her marketable skills, she had another source of income from freelance writing when her postdoc abruptly ended. She currently works in the field of medical communications. In 2019, she came across the Medical Writers Association. She had heard of medical writing but did not know how to enter the field. She began the process of exploring the possibilities, again. After studying the trends in the field, she decided to change her writing focus to healthcare. She dubbed herself a health writer on LinkedIn, leading to requests from clients for pitches. A year later, while scrolling on LinkedIn, she came across a posting for a science writer at a global biotech company, applied, and was hired. Her freelance work and self- employment stint led to the job in medical communications. Her experience confirmed her belief that a PhD trains you to learn quickly, enough to inform or educate others.

Gertrude made many lemonades of the lemons thrown her way with her ingenuity. But her experiences also instilled a sense of responsibility to support the next generation of international scientists. She creates online content to provide career development support for international trainees and openly talks about her immigration journey and her path to permanent residency. She makes it clear that it is expensive to sponsor oneself, bringing us back to the disparity of international experience. The cost can be an additional burden for those from developing countries and low-income backgrounds. As a result, she advises internationals to start exploring and planning for immigration pathways early rather than wait until graduation.

Given her experiences and reactions from her audience, Gertrude advocates for transparency and specialized support for international scientists. She remarked that it is extremely demoralizing when international scientists in the job market realize that many companies will not hire them because they lack visa sponsorship. Many international trainees openly convey to her that coming to the United States for higher education seemed like such a good option until it wasn't. The shared sentiment is, Why didn't anybody say anything about how hard the job search is? The situation is worse when your family back home has anticipated a promising future for you after obtaining a PhD in the United States, yet you're struggling to get hired.

Many international scientists feel ashamed, frustrated, and betrayed. In the age of social media, many are vocal about their experience and discourage other international scientists from pursuing higher education in the United States. Therefore, to recruit, train, and retain international scientific talent, universities must invest in specialized career and professional development support for international graduate students and postdocs.

Harnessing a Creative Mindset to Move through the Career Maze

CHAPTER TWO

Cultivating a Mindset for Lifelong Professional Development

Mindset (not intelligence or skills) is the core driver of success in the long term. It drives our relationships—with ourselves, our abilities, our communities, and our circumstances. The renowned Stanford psychologist Carol S. Dweck's scholarship highlights the importance of a growth mindset in human flourishing. A growth mindset is crucial for international scientists as we can fall prey to deficit thinking while navigating restrictive systems and barriers. This pattern further impedes success as deficit thinking limits the imagination of possibilities and results in eliminating creative options. Conversely, asset-based thinking prioritizes identifying strengths, exploring possibilities, embracing growth opportunities, and not being defined by limitations. In this chapter, I outline how to move from deficit thinking to an asset-based and growth mindset by leveraging curiosity.

Key Terms

Deficit thinking involves focusing primarily on problems or environmental barriers. For example, international scientists exhibit deficit thinking when we make career decisions through the lens of the US immigration system. In doing so, we carve out a small sliver within the defined rules of a wholesome pie. We pursue the fraction of jobs that we believe are immigration-friendly and turn a blind eye to an array of professions that we may otherwise find fulfilling. Fixed mindset is another example of deficit thinking.

Fixed mindset is the belief that our intelligence and abilities are innate and fixed. We practice a fixed mindset when our inner voice says, "I am not good at X" or "I am not gifted or talented in X." Both examples result in eliminating options. If these habits seem familiar, next time you can check yourself by naming them.

Asset-based thinking focuses on strengths and opportunities. In their book, *Change the Way You See Everything through Asset-Based Thinking*, Kathryn Cramer and Hank Wasiak encourage changing focus from problems and limitations to strengths and what's working (Cramer and Wasiak 2006). This is not the same as positive thinking. By focusing primarily on strengths and opportunities instead of weakness and constraints, we rewire our thinking to devise creative strategies to achieve success. This is akin to expanding the pie instead of playing zero-sum games, which are either dated or weren't equitable by design. For example, leading with our strengths and skills in a new environment provides the confidence and courage to acknowledge our areas of growth and work on them. In exploring careers, apply asset-based thinking by focusing on your interests and motivations to determine which jobs to pursue. Once you identify areas of interest, you can find the employers and organizations that may support your immigration needs. Asset-based thinking also prioritizes things we can control, which in turn equips us with agency amid adverse conditions and reduces self-limiting habits.

Growth mindset centers on the belief that talent and abilities can be developed through learning, effort, community, guidance, and practice (Dweck 2006). By simply adding the word *yet,* you can move from a fixed mindset to a growth mindset. For example, when you say, "I am not good at *x* yet," your next thought will be, "How can I get better?" Therefore, you change your focus to developing abilities and exploring possibilities, which are again factors you can control. In practice, a growth mindset leads to a journey of continual self-development and reinvention through lifelong learning. Scientific research training relies on and fosters a growth mindset. In research, we don't let roadblocks, dead ends, failures, or restrictions define us. We explore, learn, and develop other approaches to discover truths and create impact.

Now that you know the terms, I encourage you to repeat the following such that it becomes your mantra: *"I won't be defined by barriers and restrictions and fall prey to deficit thinking and a fixed mindset. I will strive to adopt asset-based thinking and a growth mindset."*

Through my own experiences and observing graduate students and postdocs I advised, I can vouch for the fact that a growth mindset is key to professional development. One of my graduate student mentees, Emmanuel Toroitich, employed curiosity and a growth mindset in his career. Born in Kenya, Emmanuel came to the United States for undergraduate studies. He chose chemistry as his major because he was initially interested in medical school. After receiving his undergraduate degree, he decided on graduate school over medical school because of the additional expense for international students. He debated between two options: working in industry for two years and then going to graduate school or pursuing a PhD directly. An informational interview with a friend in industry helped with the decision. He learned that the starting salary in biotech after the bachelor's degree was only marginally better than a PhD student's stipend. Also, he learned that visa and permanent residency options are better for those with PhD degrees. This led him to decide that graduate school was the best route.

As he was broadly interested in learning about the business side of science, he participated in the PhD Plus professional development programs at UVA in business skills, entrepreneurship, commercialization and venture capital, and management consulting. Even during his undergraduate years, he took economics and accounting classes. In retrospect, those undergraduate classes and PhD professional development modules were tremendously helpful in his post-PhD career trajectory.

Emmanuel also followed his curiosity to explore business of science for a post-PhD career path. He regretted that he had not started exploring possible careers earlier in his PhD training. By the time he was preparing for a career in consulting, he was less than a year from graduating, so he missed the early recruitment timelines of big consulting firms like BCG, McKinsey, and Bain. As a nimble, forward-looking systems thinker, he looked for alternate ways to enter the field. He started interviewing for niche healthcare consulting firm and equity research positions as well as related postdocs. He took an innovation postdoc, his first foray into consulting work.

Two months into the innovation postdoc, he took a full-time position as senior analyst at a healthcare consulting firm. He explained, "Although the postdoc was good, the full-time role came with double the salary and exactly what I wanted to do—healthcare consulting work." He liked the diversity and defined nature of projects, which typically lasted from six weeks to two months. He typically worked on two projects simultaneously with different teams. As a result, he learned a lot within a short period and had the benefit of learning from diverse mentors and colleagues.

Even as he enjoyed his role in the consulting world, he kept his focus on current and future workforce trends. He noticed that the consulting industry was in flux, leading to regular layoffs. He stayed ahead of the curve by planning forward. Instead of working with multiple biotech clients in a consulting firm, he decided to focus on business development in one biotech firm. He applied for business development positions in biotech, which led to his current role as strategy analyst, corporate development, at a biotech company. In hindsight, he also realized that the work-life balance is better in his current job. I was amazed at his immaculate planning and his humility.

Curiosity Is at the Center of Asset-Based Thinking
and a Growth Mindset

At this point, you may be wondering how you can apply asset-based thinking and a growth mindset in your daily professional lives. It's simple: lead with *curiosity*. Scientists are curious by nature. Curiosity drives our scientific pursuits. In research, curiosity enables having an updated understanding of scientific fields, observing patterns, and identifying directions. Curiosity also fuels scientific thinking by questioning assumptions and striving for objectivity (Majumdar 2023). For professional development and career advancement, you should apply the same curiosity to other aspects of your lives. Begin with acknowledging that *you don't know what you don't know*. Make that your second mantra. Be curious about yourself and your environment. Treat your career with the same curiosity that you bring to your research. Regularly update your knowledge; question defined norms, assumptions, and prevailing narratives about career paths; and explore unknown domains. Be curious about the current employment landscape and observe trends and future projections to visualize possibilities. Look for successful examples among international scientists who have charted the course.

HOW TO APPLY CURIOSITY IN CAREER AND PROFESSIONAL DEVELOPMENT

In this section, I discuss how to leverage curiosity to gain self-knowledge and practice systems thinking and lifelong learning. Self-knowledge, systems thinking, and lifelong learning are habits and approaches that ultimately enable asset-based thinking and a growth mindset.

Self-Knowledge

Self-knowledge involves identifying your core values, skills, and interests. Spend time reflecting on your motivations, what provides purpose and meaning, your relationship to work, and, more broadly, your civic responsibility, positionality, and vision for the future. In simple terms, what are the attributes that make you feel motivated and happy in your work life?

WHY IS SELF-KNOWLEDGE IMPORTANT? First, self-knowledge serves as a touchstone in a changing work environment and employment landscape. You will likely navigate many jobs in your career, in various fields and sectors. While approaching diverse jobs and organizations, you need to be aware of your motivations and interests to assess whether they are a good fit. Second, self-knowledge is important for honest and authentic communication with potential employers (Attig 2019). Job seekers who spend time evaluating how a job description and an organization's culture align with their values and interests can communicate their interests honestly and coherently (Attig 2017; Willis 2020). Third, self-knowledge will guide you amid restrictions. For international scientists, immigration policies can pose barriers to career advancement, so self-knowledge becomes even more important as you weigh employment options that resonate with your core values and interests as well as communicate your strengths to employers (Majumdar and Schneider 2021). Fourth, self-knowledge is an indicator of emotional intelligence; it involves assessment of your competency and strengths as well as informs your professional development goals for career advancement. Employers value individuals who possess these leadership attributes. Self-knowledge will go a long way toward helping you to advocate for yourself and collaborate with your employers to achieve professional development and advancement at different career stages (Majumdar 2021). Ultimately, self-knowledge assists in identifying your strengths and treating yourself as an asset first and then apply asset-based thinking. It will also expose your areas of growth and development, opening avenues to practice a growth mindset.

HOW CAN YOU DEVELOP SELF-KNOWLEDGE? You should engage in the self-reflection process just as you approach research: in methodical and data-driven steps (Majumdar 2021). I recommend that you follow these three steps.

- **Retrospection**, the act of reflecting on your past, can be a great first step to gaining a fuller understanding of yourself. You can identify themes and driving motivations, whether reflecting on the past year or your overall professional history, such as why you picked a specific undergraduate major, why you applied to certain graduate programs, and why you selected a specific research area. If you have prior work and volunteering experience, what motivated you to pursue those opportunities? Retrospection helps us connect the dots of a nonlinear career trajectory to create a coherent narrative of a complex career path (Ranieri 2015).

- **Introspection** involves reflecting on your current thoughts and feelings. After connecting with your past, you should examine your present self by considering the following: What parts of your day make you happy, inspired, and engaged? What motivates you during challenging days? How does your worldview influence your relationship to (and at) work? One easy exercise is to make a list of likes and dislikes in your current work to help you identify your interests. Comparing observations from retrospection and introspection will highlight your professional and personal growth and achievements, instilling confidence in charting your path forward.

- **Extrapolation** is the process of estimating the unknown by applying trends of the known. Given your past and present motivations, you can forecast your desired future. You can begin by reflecting on the following questions:

  - What kind/s of future would you like to contribute to?
  - What problems or societal issues do you want to solve?
  - What systems and practices would you like to change?
  - What are non-negotiable, core aspects of work for survival and fulfillment?
  - What are undesirable parts of work that you can tolerate in reasonable doses?

Reflecting on these questions will guide you in a process of elimination and focus on specific kinds of jobs that are good fits. Extrapolation will help you imagine different versions of your future self and inform the range of work and professional cultures that will inspire you.

There are free, online self-assessment tools for scientists offered by scientific societies (e.g., AAAS, ACS) to reflect on values, skills and interests via scoring. If self-scoring makes you uncomfortable, journaling is an open-ended approach of self-exploration where you can summarize observations from retrospection, introspection, and extrapolation. You should engage in self-reflection regularly, perhaps every three to five years, as your interests and skills will evolve. Your core values will become more consistent, and they may move you to consider a different field or a different way to contribute.

Systems Thinking

Systems thinking is a holistic and disciplined approach to examining problems or issues fully and accurately before acting (Goodman 2016; Meadows 2008). Systems thinking involves investigating structures and systems and their relationships with factors—at the micro or macro level—that contribute to trends. It requires taking a top-down view of things to determine your positionality in the grand scheme. It also allows you to ask better questions before jumping to conclusions or taking immediate actions. You should apply systems thinking in your research as well as career planning. Too often scientists focus on the minutiae of a problem without carefully assessing the landscape and influencing factors. Focusing primarily on the microlevel and missing out on macro trends is an error-prone strategy in both research design and future career planning (Majumdar 2023). Therefore, you must maintain a balance between a top-down and a bottom-up view and observe patterns and connections between influencing factors.

Leon (Jun) Tang's career from PhD to biotech venture capital followed by entrepreneurship is a model of having a bold vision and using systems thinking in career planning. Jun grew up in China in a low-income family. He was self-aware and candid that social and economic mobility were his primary drivers for a career. He mapped two potential futures, one in China and one in the United States, and compared projected estimates of salaries with a STEM graduate degree. Moving to the United States for his PhD in life sciences presented a financially stronger future, with diverse options.

Having realized the prime currencies of academia, he remarked, "From the academic point of view, publications and grants are the capital, and this is a capitalist country after all." He also investigated the business model of academic science: the scientific ecosystem and financial and operational systems. He elaborated, "Once I understood the academic science business model, the publish or perish system made sense. You can see how federal policies in the 1990s impacted how science is done now." With that understanding, he planned his strategy for success. He secured independent funding during his PhD training. With his adviser's support, he wrote a proposal for the American Heart Association's predoctoral fellowship, which, while highly competitive, was open to international students.

He had a clear vision for his career too. As he studied the systems (e.g., immigration) that had an impact on his life, he realized he would need permanent residency (via a green card) for career agency and independence. Therefore, he approached postdoc training strategically, with the primary goal of building competitive research and immigration portfolios in order to obtain a green card. After receiving his PhD degree, he did two postdocs on developing cancer immunotherapies. He succeeded in getting a green card during his postdoc after building a competitive profile. In the meantime, he kept exploring different career paths and upskilling to differentiate himself. For example, he got a proficiency certification in medical writing. He said, smiling, "You know, I am a pretty good writer, even if English isn't my first language."

Jun referred to his career development approach as "using a shotgun strategy to fulfill professional vision and purpose." He argues that passion, curiosity, financial success, and social impact are not mis-aligned values. His professional vision was to make a large-scale impact by advancing innovative oncology therapy. He concluded that the private sector, likely in the business side of science, would provide him with the right venue for scaled impact. So he joined Cancer Research Institute as senior manager of a philanthropic venture fund, where he led due diligence on investments. Following his stint in the nonprofit sector, he joined Barclays investment bank as assistant vice president, conducting equity research work. He outlined the differences between the sectors. Although he learned a lot about the venture capital field and investments at the nonprofit, it was a less intense culture than the cutthroat environment of investment banks, where mistakes can get you fired. The intense culture in banking was unsustainable, so he switched from investment banking to business development in biotech for a better work-life balance. He joined a global biopharma company as senior director of business development, where he led due diligence teams on search and evaluation and merger and acquisition deals. Harnessing his multiculturalism and goal of scaled impact, he ventured into entrepreneurship. In 2022, he founded InScienceWeTrust Community, a nonprofit enabling biotech companies to maximize their potential in the United States and Asia Pacific. Part of its mission is guiding academics in translating scientific breakthroughs to commercially successful healthcare products. Jun's story is a fitting example of how bold vision and systems thinking can expand your career possibilities in fields where international scientists are underrepresented.

HOW TO APPLY SYSTEMS THINKING IN CAREER DEVELOPMENT As evident in Emmanuel's and Jun's stories, career planning presents a practical case of systems thinking. While approaching our future, most of us typically focus on available jobs today without thinking about issues and trends that will affect the future. You can apply systems thinking by being problem focused or issue focused (instead of job focused) to develop a professional vision (Majumdar 2023). First, you should ask what societal issues are contributing to jobs growth in specific fields. For example, due to climate change, both the public and private sectors are focused on sustainability. Also, by being problem or issue focused, you are more likely to match your professional pursuits with your values and interests. For example, Jun was interested in innovative ways to scale the impact of cancer immunotherapy, which shaped his career vision. Second, visualize unrealized possibilities. In addition to knowledge of current career fields, you should learn about new and emerging fields that may propel future jobs growth (or slowdown, as Emmanuel observed). Studying and forecasting labor market trends will enable you to prepare for jobs that may exist in the future and visualize future possibilities (see reports by the Burning Glass Institute or the World Economic Forum). For example, the advancements in genomics and AI can accelerate precision medicine goals, likely expanding recruitment of computational biologists, mathematicians, and physicists in clinical fields.

Systems thinkers are better positioned to consider the socioeconomic, geopolitical, and environmental factors that can lead to innovation or disruptions in a field. If you are a scientist motivated to mitigate the climate crisis, for example, ask the following questions:

- What factors will influence innovation in a climate tech field?
- Which institutions and organizations across the public and private sectors are doing the most innovative research, designing resilient systems, and advocating for evidence-based policies?
- How are these innovative initiatives funded?
- Where are the gaps?
- What kinds of collective action or collaboration could address those gaps?
- Where can you make the most impact?

Like staying up-to-date on your research field, practice systems thinking in career development by broadening your knowledge of industry trends and influencing factors. First, get in the habit of reading outside your field; for example, follow business and economic news and policy reports related to the problem or issue of interest. In the case of climate change, you may want to follow the World Economic Forum or the Intergovernmental Panel on Climate Change (IPCC) to gauge global trends. Second, talk to professionals involved in diverse roles within the field and attend talks by experts. You have wide and open access to knowledge and expertise on a university campus, so leverage that social and intellectual capital. For instance, if you are a scientist interested in climate change, you should also attend talks in the policy department to stay current on climate policy and funding that will impact research directions.

EXERCISE   How would you apply systems thinking to your future vis-à-vis your immigration status?

Starting Prompts

· Which fields are projected to have the most growth?
· Which jobs in those fields are most likely to require scientific skills and acumen?
· Which fields suffer from a shortage of high-skilled labor and have national importance (and therefore more likely to sponsor visas)?

Lifelong Learning

To remain competitive in a dynamic world of work where knowledge is rapidly advancing, you must keep learning new concepts, techniques, and skills. You may have already spotted the parallel; you practice lifelong learning in PhD and postdoc training. Reading new scientific papers and learning new techniques and skills are required to remain relevant. Scientific training intrinsically centers inquiry-based and curiosity-driven lifelong learning.

A PhD is not the end of the road when it comes to learning and upskilling. Like research, careers now are unpredictable and nonlinear.

The majority of PhD holders regularly change jobs at all career stages (Okahana 2019). Moreover, due to technological advancement, domain or technical skills have a shorter shelf life, so scientists must apply lifelong learning habits that they exercise in research to also stay competitive in their careers.

APPROACHES TO CULTIVATING LIFELONG LEARNING HABITS   Growth mindset approaches such as self-knowledge and systems thinking will help you craft development goals each year. On the one hand, systems thinking will lead you to gather information that provides insights on the skills required to succeed in emerging fields and interesting jobs. On the other hand, through annual self-assessment, you can evaluate skills gaps and develop your knowledge and skills development goals. Based on these insights, you should attempt to develop new skills and knowledge every year, or at the very least push yourself to learn or do something new each year, which will propel your development. For example, I jumped at the prospect of writing this book, a genre that is completely new for me. But instead of feeling afraid, I was excited because I knew this presented an opportunity for professional development. A growth mindset provides the power to overcome fear and discomfort and venture into new areas.

Our employers and universities understand the importance of lifelong learning. Most employers have free learning and development courses for employees, such as management institutes and leadership training. If you are a graduate student or a postdoc in a US university, look outside your departments and labs. Attend seminars on interdisciplinary topics; audit a course; and take advantage of the professional development programs, articles, and resources afforded by the library, your graduate school, the postdoc office, the career center, and centers for teaching and learning, among others. Finally, many universities provide free access to LinkedIn Learning. Access to knowledge has never been easier, and you don't have to spend money to take advantage of it.

That said, you must invest time. Lack of time is cited as a common barrier to lifelong learning. Time is indeed scarce. However, consistency, not intensity, is the key to cultivating lifelong learning habits. Make time in your weekly schedule to consistently engage in learning. Listen to a podcast during your commute. Read articles during your coffee break or between experiments. Watch a LinkedIn Learning video every weekend. Attend talks outside your primary field once a month. Even one hour per week dedicated to gaining broad knowledge of various fields will pay dividends if done consistently over the long run.

As I elaborate in chapter 5, you don't have to do this alone. Learning and development can be a communal experience. Expand your professional network on social media and follow their posts. I learned a lot from social media posts about PhD career and professional development experts during my postdoc, which eventually enabled my entry in the field. Schedule an informational interview every few weeks to learn about a new career field. Join interest groups and forums for peer learning. Find an accountability buddy to discuss your efforts every month. Ultimately, lifelong learning comes down to managing yourself and managing your network. Cultivating asset-based thinking and a growth mindset involves cultivating long-term habits such as self-knowledge, systems thinking, and lifelong learning. These habits are incremental and favor consistency. You may feel frustrated by the minimal return on investment in the beginning. But without being aware of it, you move an inch on the path to a healthy mindset every day. Like exercise, cultivate a routine such that these adaptive behaviors become habits and mindsets.

Emmanuel and Jun highlighted the critical importance of curiosity, systems thinking, and lifelong learning when reflecting on their careers. When I asked Emmanuel what advice he would give to the next generation of international scientists, he articulated three lessons from his life. First, keep planning for the future, and do not get comfortable in an organization after working there for a few years. Especially in the current employment climate when layoffs are regular and more precarious for visa holders, one must stay ahead of the curve. Second, keep learning to continue thinking ahead. Leverage informational interviews and take advantage of university resources. Third, take a systems view of the field and sector. Regularly read business news outlets. He recalled that he had impressed his interviewer with knowledge of their latest successful venture. Similarly, Jun highlighted the critical importance of systems thinking and multicultural fluency in the global economy. He emphasized that one needs to understand the industry at both the global and local scales. Once you do, you can spot current trends and identify patterns and opportunities. He remarked, "I believe in the simple principle that there's always a choice lurking. I spot an opportunity and jump into it." And that is the very essence of an asset-based mindset enabled by curiosity and systems thinking.

An asset-based and growth mindset will provide a sense of agency in your life, especially when circumstances render you powerless. It will have an impact on your well-being in the long run. You owe it to yourself, so start today.

KEY HIGHLIGHTS

- Barriers and restrictions can result in deficit thinking and self-limiting behaviors. Moving from a deficit-based to an asset-based mindset and from a fixed to a growth mindset will give you agency in navigating your professional life and career.
- Curiosity enables a growth mindset. Practice a growth mindset by introducing *yet* in your fixed mindset voice. "I am not good at X thing *yet*. Maybe I can try Y to upskill."
- Self-knowledge, systems thinking, and lifelong learning enable professional and career success.

CHAPTER THREE

Engineering Creative Strategies
for Success

While a curiosity-driven mindset broadens our imagination, creativity—the ability to generate new ideas or multiple possibilities—gives shape to it. Creative thinking helps us navigate restrictions. It is key in problem-solving, especially in designing parallel viable approaches when a traditional path to success is obstructed. A practical analogy is taking a different route on Google Maps if the traditional route is closed or clogged with traffic.

The general fear of pursuing multiple paths is the notion of sunk cost, that is, reluctance to take a course of action because you have already invested heavily in a different one. However, it need not be a zero-sum approach. Some of the investments you make in one path are transferable to others. Creative design is the art of identifying shared investments that can be applied to multiple parallel avenues. You have prior experience as scientific research requires us to design creative solutions. When experiments fail, you troubleshoot or try parallel approaches or techniques to achieve your aim. In this chapter, I encourage you to exercise the same creative muscles in your professional life.

How can you best apply creativity to your professional development? In the previous chapter, I outlined how systems thinking will expand your awareness of factors contributing to a problem. Creativity will help you map ways of problem-solving by looking at connections between those factors. In the following sections, I highlight strategies for creative design through reverse engineering multiple paths to achieve your desired goals. Then I demonstrate how to apply the framework to two examples from your own professional life, securing funding and career planning.

CREATIVE DESIGN

Mindset Recipe

Mindset is key to success, including formulating creative strategies. To be truly creative, you need to bring self-knowledge, humility, critical thinking, imagination, comfort with uncertainty, and agility—all hallmark traits of a good scientist—to the process.

Let's begin with self-knowledge. You need to be aware of your values and interests in parallel to your assessment of skills. Self-awareness and clarity of purpose are important when defining success. Consider reflecting on the following:

- Why do you want to pursue a certain goal?
- What is the purpose?
- What constitutes milestones and metrics of success of the goal?

To honestly delineate "why," examine your core values and interests. What motivates you? Why is something important to you? Awareness of skills and interests is easier than awareness of values. However, values drive our decisions and fulfillment in the long run. So spend some time reflecting on your non-negotiable values.

Have the humility to acknowledge your ignorance about the journey you are embarking on. Repeat this often until it's a part of your psyche: *you don't know what you don't know.* Much like completing a literature review to gain a better understanding of your research field, you need to gather information for a holistic review of your goals and approaches. Background information is key to formulating viable strategies and making informed decisions. You don't have to do this alone; your mentors and advisers can and should help. The tools outlined in the center of figure 4—people, resources, information, and communities of support—will help you design parallel strategies for success.

During your holistic review process, you are bound to get conflicting information and varied viewpoints. Always vet your sources. As with research, explore diverse sources of information while being cognizant of positive bias (your own and your sources'). In general, be skeptical of anyone who sounds too assured and lays out narrow definitions of success; question norms and assumptions (your own and others'). Creative design typically leads to disruption or innovation. While it is good to be aware of norms, you do not have to be constrained by narrow boundaries, especially if they present barriers to you as an immigrant. For example, prestigious international conferences are often touted as hallmarks for academic networking. However, they can be inaccessible to international scientists for visa-related or financial reasons. Therefore, instead of draining your energy and resources pursuing such conferences, you could participate in small conferences, such as those of regional chapters, that may have higher returns on investment when it comes to building professional connections. Further, look for successful examples, particularly among the international scientific community. You are most likely to learn from them about creative ways to navigate barriers for immigrants. Do not be disheartened if you can't find successful examples and role models who share your international background. You can become that role model for others.

Imagination is key to pushing boundaries of thought and designing parallel possibilities. With clarity of purpose, definition of success, and knowledge of approaches and resources, you can start mapping three parallel paths. You may find it challenging to map additional paths beyond the traditional one (linear path 1 in figure 4) Alternatively, if you have lots of imagination, you may have lots of ideas and may struggle to prioritize the three paths. In either case, pragmatic considerations or restrictions can serve as potent drivers. In their book, *Designing Your Life,*

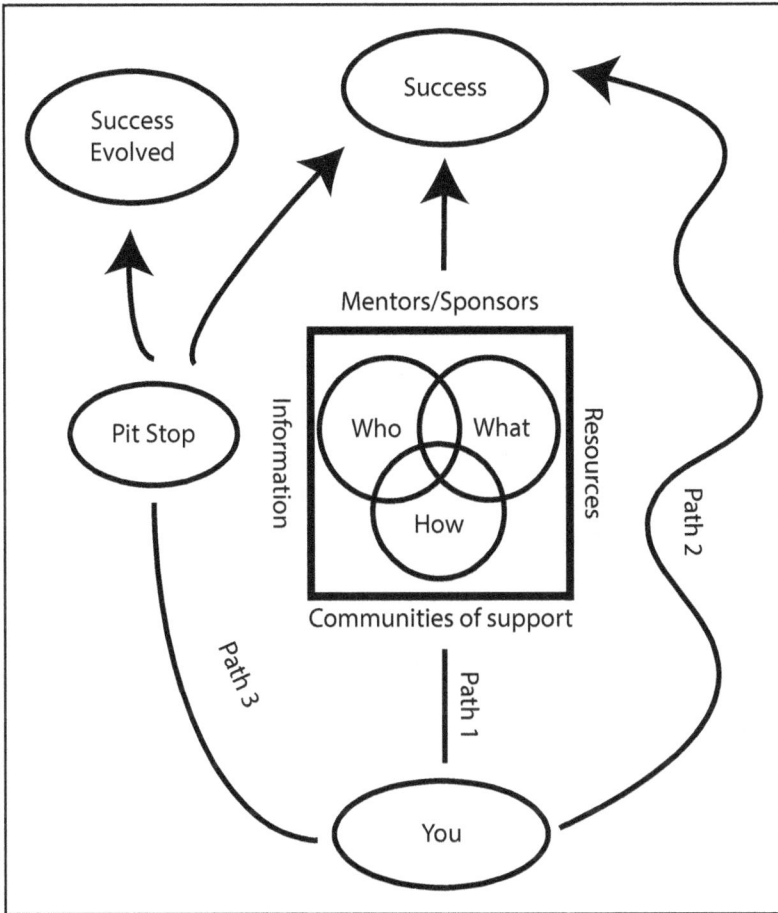

Figure 4. Framework of creative design by mapping parallel paths to success.

Bill Burnett and Dave Evans call these restrictions "gravity problems": like gravity, they are a drag and not actionable or solvable by an individual (Burnett and Evans 2016). For most international scientists, immigration policies pose such a gravity problem for access to opportunities. Identify and integrate gravity problems in your design and create an if-then logic around them. I discuss this more fully below in the sections outlining the process of designing paths and examples of application.

Finally, you will need to grapple with uncertainty and risk throughout the process of creative design. That is the essence of life as a scientist. As scientists face uncertainty in research regularly, we tend to become resilient and adaptable. Risk assessment, however, is an underappreciated skill. It is natural to fear and avoid risks. It is also easier to preach than to practice not being risk-averse, particularly while operating in restrictive systems. However, we international scientists already took a big risk when we left the comforts of our homes, our countries, and our social support systems to pursue scientific research in a foreign system. During our interview, Sina Safayi said, "We international scientists embrace disruption and change paths toward growth. Mission-driven, we regularly make sacrifices to pursue professional goals far from our home countries." That first risk of moving to a new country and starting in a new environment remains the biggest to date. It tends to get easier to take risks after that, as we gain know-how along the way. Moreover, one silver lining of surviving amid the challenges of immigrant life is our ability to exercise agility. We regularly find ourselves exploring alternatives and pivoting or adjusting our plans. As a population, we have accumulated significant muscle memory when it comes to comfort with uncertainty and being agile.

FRAMEWORK FOR ENGINEERING CREATIVE AND AGILE STRATEGIES

How should you map strategies for success? The gold standard is to start with why (purpose), then move to what (goals) and how (approaches). Devising strategies should be a collective endeavor. Your strategies for success will be influenced and informed by your loved ones and role models. It is a collaborative process, and I hope you involve your family, mentors, and communities of support. I elaborate the framework below and demonstrate how to apply the framework in ensuing examples.

1. Purpose

The key starting point for designing strategies is to define "success." What constitutes success (in a specific context) for you? Often the definition of success is tied to advancement norms in the field. It is okay to acknowledge that you are following them. However, assessing the extent to which your values and interests align with the norm is important. I say this for two primary reasons. First, when the traditional markers of success present potent barriers for international scientists, you may need to brace for an uphill battle. This would require coming up with approaches for maintaining your well-being and confidence, particularly through challenging times. At some point, if you feel exhausted and as though it is no longer worth your effort, you may need to reframe success. This will involve truly evaluating the sunk cost, which includes considerations of your (and your family's) well-being. Second, with experience, your interests (and perhaps your values) will change in the long run. Evolving interests and values may inform changes in your definition of success. This is common; definitions of success change at different stages of one's professional life. Clarity on what constitutes success for you at different stages is, therefore, important.

2. Destination

With clarity of purpose (why), you can next move to the goal(s) (what) part of the framework. When defining "what," or your specific goals, consider the following:

- What is the end goal or destination? Knowing where and when to stop is important.
- Can you visualize success? Do you have access to some models or examples of success?
- What are some conditions that need to be met?

- Where are the gaps in your knowledge that lead to assumptions? Remember, *you don't know what you don't know.* How will you find information to test your assumptions?
- How will you quantify success? Do you have short-term and long-term goals that include metrics? In chapter 6, I detail best practices of goal setting.
- What is your timeline for achieving success? What milestones will enable you to visualize progress?

3. Path(s)

Third, reverse engineer paths to success. This is driven by the central question: "What are some paths that will get you to your destination under different circumstances?" As I mentioned before, try to broadly design three parallel paths that will move you closer to the end goal.

PATH I   The first path (path 1, figure 4) is typically the traditional, ideal, and linear (perhaps the shortest) path to success. Map that first as you can visualize this path easily. In the process, recognize that you are operating under positive drivers as well as gravity problems, or restrictions. Describe the essential components, both positive drivers and barriers, in navigating path 1. The barriers will lead you to devise an if-then logic to design your parallel path 2. For example, for path 1 to materialize, you will most certainly need x. However, if x is no longer an option or presents a high barrier, how else can you get there by substituting y for x or without x?

PATH 2   You may design path 2 with the logic of avoiding the barriers in path 1 and take a longer route. You may choose to focus on meandering but incremental progress toward your destination. That approach may make the second path (path 2, figure 4) wavy, even though it gets you to your destination in the long run. International scientists regularly resort to the wavy career path. Indeed, often we do not have a choice. We take jobs that are amenable to our visa status while working on permanent residency status that opens broader and better opportunities in the long run.

Sanjana Mukherjee's journey exemplifies path 2. Born and raised in India, Sanjana aspired to become a global health scientist working at international organizations like the United Nations. She came to the United States in 2014 to complete a PhD in microbiology. During her PhD training, she improved her science communication skills by volunteering at many science outreach events. She believes that the science outreach and communication experiences helped her secure a postdoc position as an ORISE public health policy fellow at the US Food and Drug Administration (FDA) in 2019. She pointed out that contrary to most jobs at federal agencies, the ORISE fellowship is open to international PhDs. During her postdoc, she worked on a range of policy issues, leveraging her fluency with statistical methods due to her background in epidemiology.

As she wanted to venture into global health policy, she took two primary approaches. First, she followed global health security and policy experts on Twitter and learned from the discourse. Many of these experts worked with the United Nations or advised the White House during the COVID-19 pandemic. Second, she volunteered in the science diplomacy subcommittee of the National Science Policy Network. Through this opportunity, she engaged with diplomats and policy makers and learned more about science diplomacy. She also met policy makers at a Capitol Hill Day organized by the American Society for Microbiology. While she expanded her network and upskilled, it was particularly challenging to enter the global health field in the government or nonprofit sector as she was on a J-1 visa (scholar) visa. The government sector jobs were mostly for citizens and permanent residents, and nonprofits weren't open to sponsorship because of a lack of resources.

Facing roadblocks, Sanjana resorted to creative engineering to design path 2. She explored whether she could do similar work in academia, as universities are more open to sponsoring work visas (and permanent residency in the long term) for research positions. She learned that policy-focused centers at universities collaborated with the government and nonprofit sectors on global health projects that she was interested in. She took a small step on path 2 by accepting a postdoc fellowship at Georgetown University's Center for Global Health Science and Security, where she worked on pandemic preparedness issues. The ORISE policy experience helped her secure the Georgetown position. Her cumulative experiences and network also led to her next position as professional fellow in global health policy at the O'Neill Institute for National and Global Health Law. She discovered the O'Neill position through her former boss (on Twitter). Her experience confirms that people, knowledge, and imagination can help chart a new path. She was excited that she is closer to her achieving her goal as the O'Neill Institute collaborates with the United Nations by doing policy research that informs its reports. Sanjana did not let immigration hurdles stop her. She explored creative options that incrementally brought her closer to her ultimate career goal.

PATH 3 Picture this scenario: What if you didn't focus primarily on barriers and workarounds? What if you decided to take the path less traveled? What if you took a detour because you were open to serendipity and new areas? That adventurous detour could change your trajectory and lead to evolution in your definition of success. My story, in some ways, reflects path 3. After completing a PhD in biochemistry and molecular biology, I sought research jobs in biotech as I had prior work experience in the healthcare sector. However, most of my desired jobs required one to two years of post-PhD experience, so I decided to pursue a short postdoc. The short postdoc would give me time to devise a job search strategy that aligned with my immigration status. I summarized what would constitute an action plan for my wavy path 2. However, my life took a detour. Through my volunteer work at MSKCC and the NYC Postdoc Coalition, I got invested in solving a different problem: How could we make scientific training future-focused, holistic, and connected to the realities of workforce development? That led me to channel my creativity into the nascent and exciting field of PhD and postdoc career and professional development. In hindsight, this new direction provided more opportunities for innovation than the humdrum of bench work in science. Six years later, my trajectory has changed definitively. My experience with path 3 has made me more open to newer paths and unknown destinations, and my definition of success has become more fluid.

Alternatively, the detour in path 3 can still connect you to your originally intended destination of success. Sina Safayi's career exemplifies a combination of paths 2 and 3. After completing a doctorate in veterinary medicine in Iran, he earned a PhD in life sciences in Denmark. He did two postdocs in the United States with the goal of pursuing independent research as a tenure-track faculty member. During postdoc training, he gained extensive experience in teaching, mentoring, and lab management and had a strong publication and funding record. However, he was wary of becoming a "perma-doc," a permanent postdoc, vying for the limited positions in the academic job market. Also, his nonimmigrant visa status was posing a challenge to his job search in industry, so he applied for permanent residency.

During the challenging search for independent research positions in academia and industry, Sina realized his desire for other opportunities. He said, "Desiring more drives growth and disruption, making risks easier to embrace." Responding to the urge, he pursued his interest in scientific workforce development. During his postdocs, he also developed career development programs and negotiated inclusive policies for early-career scientists. His efforts led to skills and experience in academic administration, leadership, and program development. Refusing to be deterred by immigration constraints in the United States, he moved to Japan to start the postdoc office at the Okinawa Institute of Science and Technology. Around the same time, his petition for US permanent residency was approved. After a short stint in Japan, he moved back to the United States and continued to lead career and professional development for PhD students at the University of Texas (UT) MD Anderson Cancer Center, UT Health Graduate School, Emory University, and Rush University Medical Center. As an academic administrator, he developed skills and experience in business development, outreach, and strategic alliances. Meanwhile, through his work and advocacy for international scientists, he gained prominence and respect among the professional community in the National Postdoc Association and Graduate Career Consortium.

Eventually, Sina faced another crossroad in his career. His unique combination of experiences in animal, biomedical, and preclinical research; workforce development; and alliance management led to his current role as principal scientist in preclinical research at Ethicon, Johnson & Johnson. Via a meandering journey, he reached his desired, unique destination. Now he can not only undertake research but also add product development and alliance management to his portfolio. His experiences make him uniquely qualified for niche roles that can be challenging for employers to fill. His journey is a lesson that creative design requires a growth mindset, agility to adjust one's course of action with new knowledge and opportunities, and risk-taking to venture into new directions.

APPLYING THE FRAMEWORK FOR CREATIVE DESIGN

Now that you have a broad understanding of the framework for engineering creative strategies, let's look at two professional examples to apply this framework: securing funding and career advancement in diverse fields.

Example 1: Securing Funding

CLARIFY YOUR PURPOSE   As I already discussed, the first step in the framework is clarifying purpose. Why do you want to pursue a fellowship or grant? It may be to establish fundability of your independent research idea or demonstrate experience in writing successful grants. Your purpose will help you define success. For instance, securing funding (your success or destination) will establish the fundability of your research, a desirable asset for a career as a principal investigator.

IDENTIFY BARRIERS AND ASSUMPTIONS   Next, identify restrictions and assumptions. Most international scientists identify lack of fellowship and federal funding opportunities for visa holders as the primary constraint or gravity problem. To devise creative approaches, you will need to gather information on what other fellowship paths are open to international students and postdocs. For example, is there funding available from the corporate sector or foundations in your field? Applying step 2 of the framework, engage in information gathering. You could consider the following sources.

- Universities typically have databases of fellowships (e.g., Pivot) to identify sources of funding for fields that are open to international scientists.
- Talk to faculty or staff in the office of research development or corporate and foundation relations in the university.
- Identify and approach international scientists who have secured funding in your field.

DESIGN PARALLEL PATHS   With clarity of purpose and background information, you can design three parallel paths to securing funding, as follows:

1. The traditional path of applying for federal funding available to international scientists (e.g., transition awards like NIH K99) is your shortest and most direct but high risk (competitive), high reward (reputed) path.
2. Path 2 could involve writing applications for fellowships and grants from nonfederal external sources (corporate or foundation funding) or institutional (internal university) fellowships.

These nontraditional accomplishments could build a strong case not just for your next job, but your immigration portfolio for permanent residency, which would open more opportunities to path 1 in the long run.

3. If you are unable to identify a realistic path 2, your path 3 could involve exploring co-proposing a grant as a coinvestigator with your faculty adviser or principal investigator (PI). Alternatively, if the coinvestigator role is not viable, you can still take a central role in securing collaborative grants and ask the principal investigator to clarify your leadership in securing grants in letters of recommendation to prospective employers. PIs recognize the importance of fundability and are typically sympathetic to your circumstances. Walk them through your framework so they understand that you have done landscape assessment and have designed parallel paths. As responsible mentors, they would, and should, try to help. I have been privy to several conversations with international scientists who dread this conversation as they presume the worst outcome. I would suggest that you do not presume the worst outcome. If they are not supportive, there is a bigger problem. If there is a conflict of interest, you may need to find another PI or find a collaborator who may be amenable.

Example 2: Career Advancement in Diverse Fields

Career advancement amid restrictions due to visa status is the bane of existence for most international scientists, at least early in their careers. Challenges increase if you want to move into positions not directly related to your PhD or postdoc research. I have personal experience with this; it was difficult to move from RNA biology research to the PhD professional development and academic administration field. While there is a lot to be said about the need for systemic change and employer responsibilities in making systemic change (see chapter 13), I want to focus here on how you can exercise creative options amid an unfavorable immigration environment.

CLARITY OF PURPOSE   Begin again with clarity of purpose. This is harder than you may think. Reflect on the past and present to identify common threads. These threads will illuminate themes or traits that motivate you, and you can extrapolate what you would like to do in the short term (one to three years). After reflecting on my career trajectory from healthcare to scientific research, I recognized that I was attracted to start-up environments. I loved solving challenging problems and working in nonhierarchical, relatively egalitarian cultures. My interest in specific problems shifted, but I was attracted to either new emerging problems or fields open to innovation. I can also gauge my shifting interests when I start fixating on a problem or issue outside my full-time job. That's how I identify my career crossroad—my urge to take a new direction—or, as Sina said, "something more."

DEFINE SUCCESS   Once you have a broad understanding of purpose, you can delineate your destination or success in the short term. You can of course attempt to visualize long-term success, but in a dynamic employment landscape, your notion of success is bound to evolve. I typically visualize and plan for up to three years as my interests and my immigration circumstances continually evolve. Irrespective of how far ahead you plan, you may wonder how you can find the jobs or fields that manifest your idea of success? Focusing on job functions (or responsibilities) will help you consolidate your idea of success and map parallel paths. That is because job functions, not job titles or industry sectors, are more likely to align with your interests and skills. Practical considerations such as location, salary, family needs certainly matter and should inform your direction. However, your job functions will be central to your professional fulfillment and personal and professional well-being.

DESIGN PARALLEL PATHS   By centering your interests and focusing on job functions, you can map creative professional paths in different permutations and combinations (Majumdar 2023). One way is to be issue or mission focused. Let's say you care about solving the climate crisis. There are myriad climate-related jobs across the public and private sectors. Start with the job function that best matches your interest and skills—say, data analytics. Now look for examples in different sectors: What are the characteristics of jobs in climate modeling across three sectors, academia, business, and policy? In this case, the sectors can serve as the focal point of your three potential paths to map.

For another design approach, consider this: Can you design your own route by mixing components of each path? You can design your ideal job by identifying three specific functions that make it ideal. Then reverse engineer three parallel paths, where each path is a job that combines two of your favorite functions, with a third mundane or suboptimal responsibility.

While mapping three career paths, reflect on the following:

· What are the short-term and long-term positive directions of—or barriers to—each path?
· What is the typical organizational culture in each sector?
· Which paths connect better with each other and enable switching from one to the other?

These creative approaches are particularly important for us international scientists who face barriers due to visa sponsorship. The above exercises will help you design paths and explore industries where you can continue to apply your desired job function and solve problems that you care about but with an employer that will sponsor work visas or employment-based green cards. As outlined in path 3 of the framework, you can create a career pit stop—a practical, short-term strategy—while you work on your immigration portfolio, which will open access to a wide variety of jobs in the long term. Alternatively, you can explore other countries with better immigration systems, providing access to similar jobs. Ultimately, to make informed decisions, you should compare the strengths, barriers, risks, and viability of each path in the short and long term, as well as the sunk cost of switching paths.

No matter which paths you choose, you will need to advocate for yourself. I speak from experience that this is particularly true if you venture into a field not directly related to your prior research experience. Unfortunately, employers remain risk averse or uninformed, and in some cases misinformed, regarding paths for employment-related immigration. In such cases, you must take leadership in showing them a viable path and collaborating with your hiring manager to make the case for hiring you.

How do you gather credible information about viable strategies? Like the funding examples, explore government or credible (vetted) databases of employers that sponsor visas (e.g., H-1B jobs), identify and approach successful international scientists, and consult with an independent immigration lawyer. Finally, leverage the power of the international scientific community to crowdsource information and knowledge transfer. Communities of support such as STEMPeers, INet-NYC, and Grad Grid rely on near-peer mentorship and knowledge sharing for professional and career development. I discuss details on learning from professional communities in chapter 5.

RECOMMENDATIONS FOR DOWNSTREAM ACTION PLANS

Based on your paths, you will need to create action plans, timelines, and a list of skills required to achieve short-term and long-term goals and milestones to track progress. Again, collaborate with your mentors and colleagues during this process. The international scientist role models can also serve as mentors. I repeat, cultivate a community of support with international peers and collaborate on knowledge sharing at each step of the framework. Also, remember to celebrate your milestones. Affirming your progress is important for your mental well-being through such challenging situations. Finally—and I can't emphasize this enough—please pay it forward. Talk about your journey so that you become a role model, a success story, and a potential mentor for other international scientists. Harness the power of the collective international scientific community to enable and celebrate one another's success.

KEY HIGHLIGHTS

- Design parallel paths to navigate restrictions and obstructions.
- Use the framework to creatively engineer multiple paths in the following steps: (1) clarify your purpose; (2) define success and your goals; (3) identify restrictions and assumptions (remember, you don't know what you don't know); and (4) leverage information, resources, and people to learn about options.
- Design three parallel paths: ideal path (path 1), wavy path with incremental progress (path 2), and adventurous and exploratory detour path (path 3).

Charting a Path with Confidence, Purpose, and Support

CHAPTER FOUR

Strengths of International Scientists

Beyond Technical Skills

I was a bit of an outlier in my family. An introvert who preferred books to socializing, I lived in my mind and a world of imagination and curiosity. I became fascinated with viruses in high school and decided to pursue microbiology for undergraduate studies. Although my parents loved the idea of at least one of their offspring showing interest in the sciences, they had never heard of microbiology or molecular biology. They advised me to consider medical school instead. I told them I would explore both options. So I prepared for the Indian medical school entrance exam while also exploring colleges with microbiology majors. As the medical school admission results were announced after college admissions for basic science programs, instead of waiting, I accepted the admission offer of one of the few colleges that offered microbiology at the time. I was so excited that I didn't want to know whether I was admitted to a medical school, so I hid my test identification number before leaving for college. To date, neither my family nor I know if I got into a medical school.

During PhD training, I recounted this story to a keynote speaker at a group dinner for the department's symposium. He commented that it was a remarkable show of resolve and single-minded determination for a young adult and welcomed me to contact him if I wished to do a postdoc. That was a defining moment for me as an immigrant scientist socialized to focus primarily on honing technical skills. I witnessed firsthand the power of lived experiences that convey human skills (typically called soft skills). Before that moment, I would wave off compliments on my cultural intelligence or adaptability because I thought that surely everyone has those strengths. But over time, I have come to appreciate that the combination of strengths and attributes from the diverse lived experiences of international scientists is unique.

Let this example be a reminder, fellow international scientist: we are more than our domain expertise (Majumdar and Schneider 2021). We make bold life decisions, take major leaps to follow our instincts and dreams, build new lives at the intersection of personal and professional identities, and form international networks through collaboration (Scellato, Franzoni, and Stephan 2012). I urge you to share your stories, to communicate your exceptional lived experiences, which enrich your workplace. Contrary to our penchant for data in scientific research, human beings typically remember and connect better to stories. That's why you should leverage the power of storytelling to narrate your strengths. Moreover, acknowledging and evaluating such strengths will also inspire you to broaden your perceived personal and professional realm of possibilities and embolden you to continue exercising the power of choice amid constraints. I speak from firsthand experience that our potent stories that distinguish us beyond our technical expertise draw employers' admiration and can open new opportunities.

IMPORTANCE OF HIGHLIGHTING DIVERSE STRENGTHS AND SKILLS

The mainstream narrative lauds our work ethic and utility. While these qualities are admirable, we are much more than our relentless work ethic. Our global outlook, multicultural fluency, and leadership qualities also demand emphasis. Therefore, we must own our narratives to highlight and amplify our diverse strengths and skills. If you are skeptical about whether this is important for career advancement, consider the following. Technical skills will get you the first job, but human skills such as leadership, communication, and cultural intelligence will advance your career in the long term. Moreover, when employers evaluate candidates with comparable technical prowess, human skills serve as distinguishing factors and additional criteria to assess your fit with the team and organizational culture. Employers also want to evaluate how you might navigate challenging situations when the chips are down. Most scientists develop professional resilience as we face frequent roadblocks or setbacks in research. How then will you distinguish your unique strengths among a group of exceptional individuals? This is where your lived experiences can help.

The lived experiences and bold choices of international scientists reflect diverse strengths and skills for professional success, such as leadership, agility and adaptability, humility, a nuanced understanding of multiculturalism, dynamic communication styles, and more. Among other attributes, you can demonstrate courage, innovation, curiosity, a sense of adventure, and risk tolerance through your lived experiences. First, our circumstances as internationals compel us to be resourceful amid constraints. We are not limited by restrictions; we find a way to overcome them. In fact, our bold life choices define us within and beyond our circumstances. For example, a scientist who is a non-native English speaker works twice as hard to develop fluency in English for scientific communication and navigates life in American society where fluency in English is somehow equated with intelligence (Amano et al. 2023). Second, our varied identities, perspectives, immigration status, and backgrounds add complexity and nuances to our experiences. International scientists have a tapestry of lived experiences because of the intersectional complexities of our circumstances and backgrounds. Therefore, we enrich professional environments with our diverse and global outlooks and multicultural values to create inclusive professional environments. Third and finally, when we celebrate our choices and lived experiences, we understand our varied strengths, which gives us agency and confidence to broaden possibilities. These are classic traits of asset-based thinking and creative engineering.

DRAWING FROM LIFE EXPERIENCES TO COMMUNICATE STRENGTHS

Let me highlight how you may identify and communicate skills and strengths in cultural intelligence, communication, courage, adaptability, and resilience.

CULTURAL INTELLIGENCE   International scientists move between countries and continents to pursue knowledge and careers, strengthening our cultural intelligence. Cultivating cultural intelligence involves motivation to learn and understand new cultures, as well as mindfulness to navigate distinct cultural scenarios. I grew up in India, a diverse, multicultural country. I lived in three culturally distinct cities in the north, east, and south. I often felt like an outsider, which motivated me to analyze norms, similarities, and distinctions in those multicultural societies. I did the same while pursuing PhD and postdoc training in the United States. In the past seventeen years, I have lived in four states in different parts of the country: the Northeast, the mid-Atlantic, and the Southeast. My curiosity to learn, observe similarities and distinctions, and draw connections between cultures has enriched my experience and facilitated my personal growth. This strength also helped as I transitioned to higher education administration, which differs culturally from scientific research.

Let me be explicit: adapting to cultures does not necessarily mean assimilating. It is more about connecting with people through similarities and differences to achieve mutual goals. My perspectives as a cultural outlier continue to add diversity and depth to my professional environments. As I mentioned earlier, I felt like an outsider even in my country of origin as I moved between culturally diverse regions. I developed a strange sense of comfort with coexisting in diverse environments as an outsider, or more accurately, a non-native, so moving to the United States wasn't challenging. I habitually ask questions (about history, purpose, norms) to learn about my immediate environments and decipher my positionality. Sometimes these fundamental questions also provide a space for natives and leaders to reflect on purpose, equity and access, and cultural evolution. This exercise also helps me assess whether the enforcers of norms are thoughtful and inclusive leaders or simply operate from a sense of authority, power, and privilege.

I outlined Leon "Jun" Tang's bold career journey in chapter 2. Jun highlighted how his multicultural fluency as a Chinese scientist in the United States provided a competitive edge. He explained:

> Typically, people think being ethnic Chinese (or broadly foreign) is a big disadvantage, right? Which oftentimes is true, especially in business fields. You have less social capital than the Ivy League types, for instance. But I had a unique advantage due to the global nature of the life science industry and China doing particularly well. I had the combination of biomedical science expertise, business knowledge, and nuanced cultural understanding of the US and China. I also understand ethnic or cultural variations between China and Taiwan to degrees that even a first-generation Asian American won't. Business is about understanding people and cultures. Science can convince people to a certain extent, but interpersonal trust is key to decision making. Therefore, I can leverage my multicultural skills to thrive in global markets.

Jun's comments aptly demonstrate the value of multicultural fluency in the workplace that an international upbringing brings.

COMMUNICATION   Living in multicultural and multilingual societies affords international scientists a dynamic relationship with languages and a nuanced appreciation of diverse communication styles—another valuable skill in a global workforce. Many international folks are bilingual or multilingual. Like many Indians, I am multilingual. I am fluent in English, Hindi, and Bengali. In addition, I acquired basic communication skills in the regional languages typically used in the Indian cities where I lived. Similarly, during my short trips abroad, I find a way to communicate with local vendors and shopkeepers. I am relatively comfortable communicating through nonverbal cues or using technology (e.g., Google Translate) in the absence of a common language.

Second, international scientists regularly code switch, not just between languages, but between communication styles and norms. Like actors, we are shape shifters, deftly moving between different mannerisms and normative styles of communication. These nimble communication traits have certainly helped me navigate oral and written communication styles as well as norms of professional presence in different fields—whether in scientific research, clinical labs, or academic administration. In graduate school, my adviser commented that I wrote my research proposal like a literature major, not like a scientist. With his help, I learned norms of scientific writing. Now, in the field of career and professional development, I switch between sounding like an academic, educator and administrator depending on audience. I have also advised graduate students and postdocs, and collaborated with faculty in STEM, humanities, and social sciences. Flexibility with a broad array of communication styles has helped me connect with individuals who express themselves in varied ways. Over the years, I documented these observations in a journal that highlighted my personal growth and my strengths in cultural intelligence and communication.

CREATIVITY Creative pursuits can extend into your scientific life. Jenny Schneider, a fellow international scientist and my friend from postdoc days, applied creativity and the diverse skills gained as a dancer to her scientific life. Jenny trained as a ballet dancer before her career in science and drew qualities of her stage persona as a dancer into science communication and presentation, for example, improvisation and public speaking. Specifically, she found authentic and accessible ways to communicate and connect with a variety of audiences. She remarked, "You are on a stage when you're presenting, even in lab meetings. Stage presence was ingrained in me as I have been on a stage since I was three years old. I also understand nonverbal communication, which I first applied in performance art and then with scientists." Jenny's confidence with creative expression and nonverbal communication also assisted in effective communication with people from many different cultures and countries.

We all have hobbies and interests besides science. Reflect on how you can apply the additional skills that you gained from your creative pursuits in your professional lives. Often they transcend culture, language, and discipline, adding dimension to your tool kit of strengths.

COURAGE   We international scientists exercise courage and take risks when making life and career decisions, sometimes while overcoming significant circumstances. This proactive courage inspires resilience in the face of adverse situations in a new environment. Exploring and preparing for careers beyond the bench after decades in the field of RNA biology was challenging for me because of the specific stipulations for nonimmigrant visas in the United States. I found very few international scientists who had transitioned successfully from scientific research to higher education administration. The fraught immigration climate of 2018 added to the challenges. I almost gave in to despair, but my career advisers told me to draw confidence from my past professional decisions and lived experiences. I reflected on my courage in two defining moments: when I confounded expectations by pursuing microbiology over medical school and when I took the risk of leaving a stable health-care job in India for PhD training in the United States. Through the PhD training, I gained skills in the process of creative engineering through roadblocks. Channeling my determination (or bullishness, as my mother would call it), I prepared for an uphill battle yet again, this time against structural barriers.

ADVENTURE   This brings me to our next superpower, *sense of adventure*. We move to a new country and embrace new systems, environments, and cultures to follow our curiosity and love of science. Looking globally, few people voluntarily leave the comforts of home. It takes enormous courage, a sense of adventure, and a healthy dose of risk taking to move from one country or continent to another. It is unfortunate, but navigating the power-asymmetric academic culture and restrictive immigration system in the United States can often dilute our sense of adventure. When I rediscovered the sense of adventure to imagine a life beyond the bench, I found myself asking again, "What do I really like doing?," "Where can I do that kind of work?," and, ultimately, "Who do I aspire to be?" Countries didn't matter; after all, if I could move to another continent once, I could do it again. I acknowledge the liberty I have, as I do not have dependents. It is harder when your decisions affect a spouse, children, parents, or other loved ones. But in all circumstances, I believe we have some discretion, or a window of choice that we may hesitate to explore. I encourage you to reflect on your past selves and explore the questions that inspired your journey so far. In other words, what would your younger, more uninhibited version want to explore? Is there a realistic path where you can marry your sense of adventure and responsibilities? Resort to the creative engineering approaches in chapter 3. Give yourself and your loved ones the permission to imagine, to dream of venturing into new domains. I assure you through firsthand experience, it feels liberating to break the shackles of restrictive circumstances, channel your sense of adventure, and manifest radical choices.

AGILITY AND ADAPTABILITY   We regularly adapt to new systems in the United States, whether it's banking and financial systems, credit histories, the convoluted healthcare system, or the ever-changing immigration goalposts. We navigate hidden rules and systems replete with foreign logic and language with agility and precision to not just cope but thrive in a competitive society. And then there's *resilience*. Much is said about the remarkable resilience of immigrants. Studies suggest that immigrants fare better than US citizens when it comes to overcoming adversities because resilience is our daily diet and exercise regimen (Kuper and Turanovic 2021). I wish this was not the case, and I don't take pride in it. An inclusive society would not put us through the wringer on a regular basis. In the meantime, we show up and excel professionally with a poker face while enduring suffering due to travel bans, visa-related existential gymnastics, financial burdens, geopolitical conflicts, fewer opportunities, distance from loved ones on joyous or sad occasions, and any number of other adverse circumstances. Should we flex our resilience muscles that are so valued by our workplaces while pushing our bodies and minds to weather hardships regularly? I leave it to you to celebrate or feel conflicted. Either way, being silent or hiding the dark underbelly of resilience does us no good. Whether you celebrate it or not, please don't shy away from voicing the collective toll of such resilience on the international scientific community.

IDENTIFYING AND COMMUNICATING PROFESSIONAL SKILLS

In addition to your lived experiences, you should reflect broadly on your professional skills. PhD and postdoc training equips you with skills beyond research, teaching, and scientific communication. You gain skills in project management, leadership, mentoring, time management, collaboration, and finance management, among others. You may not be fluent in the language of management, but you manage your research projects, mediate collaborations and partnerships, and mentor other scientists in training. Storytelling scaffolds can help you identify these skills as well as communicate them. One such scaffold is STAR, the acronym for Situation, Task, Action, and Result. You can also add another "R" for reflection to make it STAR(R).

Situation: Background and setting the stage

Task: Goal or challenge

Action: Your position or work

Result: Outcomes

Reflection: Lessons learned

Broadly speaking, the sections of a scientific manuscript also apply this scaffold.

Situation: Introduction

Task: Research question(s) and experimental design

Action: Experimental methods

Result: Results

Reflection: Discussion

Going forward you can use this scaffold to convey your professional skills. Think of something interesting that you did or experienced this week and write it using this scaffold. Then identify any skills or strengths that the narrative conveys. Any story could point to more than one skill or strength. But if you would like to highlight one of them, you may need to reframe the story to draw the central focus to the specific skill. Alternatively, you can take a specific skills type and scaffold an experience to highlight it.

Examples of STAR

In the following example, I communicate how scientists create efficiencies in lab budgets through effective use of consumables.

SITUATION   As a molecular biologist, I used lots of pipette tips for experiments. Pipette tips are the most frequently used consumable in molecular biology labs. Competing vendors would leave free samples of their latest pipette tips. I played around with different samples to get a better sense of which tips were good for what kind of experiments.

TASK   When I discussed this in group meetings, we quickly realized we had been using unnecessarily expensive tips. We needed to devise an economical approach to our workflow.

ACTION   A group mate and I assessed the frequency of different experiments in the lab. We came up with a plan for how many kinds of pipette tips to order and from which vendors and allocated duties to lab members who were in charge of ordering and tracking usage.

RESULT   Over a year, we ended up saving a percentage of our consumables budget simply by tweaking our use of pipette tips.

I use the above story to highlight skills in managing lab finances. I could reframe it to highlight leadership or collaborative problem solving by focusing on how I identified, approached, and solved the issue. These are mundane, everyday lab experiences that we discount as scientists. But these experiences highlight professional skills valued beyond academia.

Like any art form, you will get better at storytelling with practice, so start a STAR journal in which you create a STAR-scaffolded narrative of weekly experiences and synthesize skills. A STAR journal can help you prepare for interviews. In most interviews, employers ask a set of behavioral questions to assess a potential colleague's collegiality and fit with the team and the organization's culture. For example, I used the following STAR story to respond to the question, "How do you approach or manage conflicts?"

REFLECTION   I take a personalized approach to managing conflict by understanding the interests of the other individual and taking their personality or preference into consideration. Let me give my mentorship example with an undergraduate.

SITUATION   During my scientific research life, I had the opportunity to mentor many undergraduate students, typically for two years.

TASK   While mentoring a new (and exceptional) undergraduate, I realized within a few weeks that she had trust issues.

ACTION   Instead of telling her what to do, I provided guidance on research design. When she presented alternate arguments or rationales, I encouraged her to proceed with her line of thinking. If the experi-

ment failed, I helped her identify the errors in design and suggest trying my approach. In this scenario, she would learn from and eventually
come to respect my experience. If she succeeded, I would learn from
her. This egalitarian environment of openness to learn and humility
built a foundation of mutual respect and trust between us.

RESULT   We learned from each other's strengths. In a year, she
learned, thrived, and even mentored a new undergraduate researcher.
Due to her significant contribution to the research project, she was one
of the authors of our research paper. Following undergraduate studies,
she joined a renowned graduate program and recently defended her
PhD, which she invited me to attend.

I imagine you have had similar experiences that you shrugged off as
mundane and unremarkable. But these common stories uncover your
mindset, skills, and approach to problem solving. Therefore, do not
hesitate to own your narratives and tell your stories. You will notice
that successful people regularly rely on storytelling to engage attention
and advance careers. By speaking up, you also amplify diverse voices
and experiences in the professional world. Despite a diverse global
workforce, only a narrow set of people are considered leaders. For all
the books, TED talks, and snappy quotes on leadership, the dominant
traits of leaders include confidence and eloquence. Wouldn't it be nice
if our stories of consistency, hard-earned competence, and collective
empowerment broadened the defining traits of leadership? Rather than
aspire to be another overconfident and entitled leader, let's truly diver-
sify the meaning of leadership. Let our stories shape cultures and systems to include international perspectives and wisdom.

KEY HIGHLIGHTS

- Your lived experience as an international scientist gives you diverse strengths beyond technical skills.
- The strengths derived from your lived experience in combination with technical expertise distinguishes you in the workplace.
- Storytelling is useful to identify and communicate skills and strengths. Use storytelling scaffold STAR(R) (situation, task, action, result, and reflection) to craft stories.
- Develop a STAR(R) journal to build your collection of skills stories.

CHAPTER FIVE

Building Communities of Mentors and Champions

International scientists travel far, leaving homes, friends, families, and social support systems for our love of research. We make fresh starts in new environments with little social capital and must rebuild professional networks and communities for support, well-being, and success. Networking may feel daunting because of your lack of nuanced understanding of the norms and know-hows of professional culture in the United States. Moreover, the process of networking and the concept of personal branding may feel inauthentic and artificial, but it doesn't have to be. Here I outline approaches for international scientists to build professional networks and diverse communities of support with intention and authenticity.

MEANINGFUL BONDS

One of the best features of PhD and postdoc training is the vibrant international scientific community. Scientific research is a global team sport. We work with curious and passionate people from all over the world who bring with them diverse stories and experiences. Amid the long and rigorous journey of scientific training, lifelong friendships are forged among individuals who share successes, failures, and, unfortunately, traumas. My ride-and-die best friend, Caryn Hale, was my bay mate in the lab and peer mentor during my PhD training. We have been each other's support system for seventeen years through creative ideas, failed experiments, challenging manuscripts, research and life rollercoasters, science outreach, exposing each other to Indian and Cajun food, career and immigration uncertainties, navigating professional life as women in STEM, and personal loss. As grueling as research life can be, it bonds people in deep and meaningful ways. I hope you stay in touch with the communities formed during graduate school and postdoc training.

Beyond departments, university ecosystems provide opportunities for community building with our international colleagues. In graduate school, I relished the dishes from different parts of the world at food festivals hosted by the international student center. Communal cooking with other Indian students morphed into long conversations and bonding experiences. During my postdoc, happy hour was equal parts sharing research ideas and complications and learning about the experiences of people from distant corners of the world. World Cup soccer was a grand affair, with celebrations, rivalries, and brackets. And there was always food: friendships were forged over expeditions in search of the best biryani in New York.

Social media has expanded the foundations of the PhD community. On LinkedIn, PhDs increasingly advocate for and support each other through career exploration in and outside academia. The international scientific community also finds ways to support and champion each other. During the peak of COVID-19 in India, Indian scientists and public health professionals in the United States and Canada responded to a call for action on Twitter and marshaled to provide credible information, resources, and medical supplies back home through a volunteer organization called India COVID SOS. The community of volunteers also helped reduce the feelings of helplessness from being far away at a time when our families and communities were suffering. We went from strangers to allies supporting each other amid anxiety and grief.

BUILDING PROFESSIONAL NETWORKS
WITH INTENTION

Whether searching for a job or exploring career paths, you should approach building networks purposefully and methodically. First, self-knowledge and awareness of your likes, interests and needs will influence who you connect with. Second, set separate professional development goals based on your needs: what you want to learn and achieve and to what end (e.g., learn about career X, improve written communication skills, receive peer review of draft manuscripts) and who can support your goals. Third, based on your goals, seek diverse networks. One person can't help you with all your goals and provide psychosocial and career support. That person may not share your background, identities, or experiences so as to provide contextual support. It is not fair to put so much pressure on any individual. It is better to identify different groups of people who can serve as adviser, mentor, sponsor, or support system.

DISTINCTIONS BETWEEN ADVISER, MENTOR,
AND SPONSOR

Advisers (e.g., dissertation adviser or research supervisor) will guide you regarding specific knowledge, techniques, or resources. They share their expertise but typically cannot take your needs, circumstances, and interests into account. For example, your dissertation adviser may teach you effective presentation skills but may not understand the specific challenges that non-native English speakers face when communicating science. Mentors, on the other hand, provide psychosocial and career guidance through some shared understanding or experience of your background, contexts, and circumstances (Mentor Doctor 2002). They can be in (or outside) your organization or be your role models. Good mentors will guide you in how to think, not tell you what to do. For example, another international scientist who has faced similar immigration challenges can mentor you on career development. Finally, a sponsor is someone who will advocate for you and enable access to opportunities. These are typically people who have the authority and ability to open doors for you. For example, a sponsor will serve as a reference or advocate for a job or leadership opportunity.

With regard to the peer review process, for example, an adviser will teach you how to do peer reviews of manuscripts in specific fields; a mentor will guide you through the hidden curriculum such as traps that first gen reviewers fall into; and a sponsor will provide opportunities for you to do peer reviews of articles, perhaps because the sponsor is a journal editor. A sponsor might also advocate for you to fellow editors to expand your access to similar opportunities. These relationships can be intersectional and exist in a continuum. A mentor can become your sponsor over time. Sponsorship is built on trust. For someone to use their reputation to advance your career, they need to know and trust you. Recognize that a mentor may not have the ability or power to sponsor you, much like an adviser may not understand your specific needs due to your background.

You should build a team of advisers, mentors, and sponsors who can support your distinct goals. As Leonard Cassuto and Robert Weisbuch describe in their book, *The New PhD,* imagine you are the CEO of your professional life (Cassuto and Weisbuch 2021). How would you cultivate your board of advisers? Consider a framework that encourages goals-oriented network building. The National Center for Faculty Development and Diversity's "Mentoring Map" offers such a network-building framework (NCFDD n.d.). The map anchors you in the center and your goals in the inner orbit. Your specific goals then inform your circle of advisers, mentors, and sponsors. You should not approach building networks alone. Consult with your advisers in the departments and university offices to identify people who can support your goals and cultivate and expand your network.

You can take a combination of active approaches (informational interviews, events, and conferences) and passive approaches (following on social media) to build networks.

APPROACHES TO BUILDING NETWORKS
Using Social Media: A Case for LinkedIn

Among social media platforms, LinkedIn is the most robust platform for building diverse professional networks. While there is a waning academic community on X (formerly Twitter) and growing ones on BlueSky and Mastodon, LinkedIn is powerful for building professional networks for diverse careers, especially for nonfaculty positions. This is largely by design as LinkedIn, in comparison to other social media platforms, still prioritizes social networking (connecting with people) over social media (broadcasting information) features (Froelich 2020). Also, LinkedIn is one of the largest job-sharing platforms, and most employers advertise job postings and solicit applications there. Therefore, recruiters and hiring managers rely on LinkedIn for headhunting and talent recruitment.

While content creators, influencers, and AI can complicate engagement on all social media platforms, it is comparatively easier to be intentional on LinkedIn. I offer a few fundamental tips to effectively use LinkedIn in the following paragraphs. For in-depth guidance and strategies, I recommend reading *Linked: Conquer LinkedIn. Get Your Dream Job. Own Your Future* (Garriott, Schifeling, and Pollak 2022).

- **Identify connections and interests on LinkedIn.** You can search for many features such as people, jobs, and learning resources by simply adding keywords on LinkedIn's search tab (e.g., science policy). If searching for multiple criteria, you can use Boolean search features. You can also search for alumni of your university who are working in your career of interest. You can follow company pages on LinkedIn for updated news, to identify employees with a shared background or institutional affiliations, and access job postings, among other features. LinkedIn groups can also connect you to professionals in your career of interest. There are LinkedIn groups primarily catering to international scientists such as GradGrid and STEMPeers.

- **Connect with people.** Use the "add a note" feature to provide the most relevant context and explain why you are interested in connecting with the individuals you have selected. You can also add a sentence about how you found out about them and any shared connections or affiliations. Try to resist talking about your research. Ultimately, the note should help them understand why you want to connect. Include your research field if it is central to forging that connection, such as research collaboration. If you are nervous about reaching out to people, especially those with a large number of followers, you can start by following them instead of connecting.

- **Lifelong learning.** You can absorb information passively on LinkedIn. LinkedIn Learning is excellent for professional development as well as lifelong learning. Many US universities provide free access to LinkedIn Learning via institutional email. You can highlight your LinkedIn Learning badges and certificates under the "License and Certification" section of LinkedIn to demonstrate your interest in learning beyond formal education and training. In addition, make a habit of reading and engaging with posts to learn about new fields or important issues. For example, I learned a lot about the trends and future directions of the emerging graduate-level career and professional development field. Through LinkedIn (and Twitter) posts, I learned about innovative initiatives, general discourse, and scholarly papers in the field. Follow interesting and credible people, and read and like their posts. Over time, try to leave a comment.

- **Build a LinkedIn profile.** Think of your LinkedIn profile as your dynamic professional panorama. In addition to building networks, think of how you would like to be known by prospective employers. As you can apply for jobs via LinkedIn, your LinkedIn profile can be used as input to create a candidate profile. Therefore, your LinkedIn profile literally serves as an online resume. It is common practice for hiring teams to look up your LinkedIn profile in addition to reviewing your application materials.

Tips for creating a good profile:

- Always upload a professional-looking photo.
- Use the "About" feature to create a narrative that highlights your professional vision and values and include your training, skills, experiences, and aspirations as supporting evidence.
- You can use the "Featured" section to highlight any news articles about you or your work or articles you published.
- Have a complete history of your professional experience that includes job titles and a summary of responsibilities and accomplishments.
- Include your service and leadership work under the "Volunteering" section. For international scientists, it is prudent in your LinkedIn profile to separate volunteer experiences from your work experiences due to visa stipulations. As visa applications increasingly ask for social media handles, take precautions to accurately reflect your work history.
- Include your complete educational history. The algorithm uses university and organizational affiliations you include in the education and experience sections to identify others with shared affiliations. Therefore, it is important that you include all institutions you have been associated with to cast a wider net(work).
- Add relevant skills, and prioritize your top skills based on your career interest. For instance, during graduate school and postdoc training, I highlighted my technical and scientific skills. As I changed career fields, I prioritized my leadership, management, and communication skills as they aligned better with higher education administration positions.

- Under "Publications," you can add your scientific publications, as well as other forms of writing such as articles and blogs.
- Under "Languages," highlight your fluency in several languages. In a global workforce, this is an asset.
- You can ask your colleagues and collaborators for recommendations. You should also offer to write recommendations for peers.

As a rule of thumb, learn from your LinkedIn role models. Make a note of what you like in their profiles and try to emulate their practices. The same goes for posts; if you like specific kinds of posts, consider what is good about them and attempt to incorporate it into your own. If you are uncomfortable writing posts, unpack what it is that makes you uncomfortable. If you don't like writing about yourself, start with writing about your field or your research in plain and accessible language, and share articles and papers that you find interesting. Engage by writing specific and helpful comments on others' posts. Most LinkedIn users do not engage actively via posts or comments, which means posting on LinkedIn will amplify your online presence and give you an edge.

Informational Interview: Active Learning and Forging Connections

Informational interviews are an effective way of learning from the knowledge and experience of others. As the term suggests, you interview someone for information about their experiences, jobs, career fields, skills for success, growth opportunities, and additional connections. Most scientists try to honor requests for informational interviews as credible and specific information is scarce and they sincerely want to help the next generation. That said, only ask people for information, not jobs. Approach informational interviews as an opportunity for learning and growth, not a transaction. The following steps can help you effectively approach and manage informational interviews.

1. *Identify people.* Based on your goals, make a list of people who can give advice or information. These people can have shared educational background, identity, or experiences and can be in fields and positions you aspire to. Start with those who have mutual connections, for example, alumni from your lab or those who went to the same university for undergraduate or graduate school. In each informational interview, you can ask about other individuals they would recommend you connect with. That's how you can start building networks.

2. *Contact them.* You can contact the people you identified via email or LinkedIn. In either format, write a concise note, two or three sentences, describing your background, your interests, and why you would like to connect with them. Do not go into a long explanation of your research, especially if it isn't relevant to your purpose for contacting them. Most people are busy, so it is important to respect their time. In an email, you can be direct and request a twenty- to thirty-minute informational interview by phone, on Zoom, or in person. To make it easy, you can even propose a few times in the coming weeks for a conversation or request alternative times. On LinkedIn, once someone accepts your request to connect, you can follow up with a request for an informational interview. Don't lose hope if you don't hear back. Follow up in two or three weeks as it's likely they were busy and forgot to respond.

3. *Prepare before the conversation.* Do some background research on them and their employer; do not ask them about things that you can easily find on their LinkedIn or public pages. Make a list of questions that align with your learning goals. Outline a few ideal outcomes from each conversation such as one new insight or piece of information or a new connection. Also, prepare a brief introduction reminding them of your background and the purpose of this conversation. The introduction will help manage expectations and direct the conversation.

4. *Conduct the interview.* Be respectful of people's time. Typically, you can reasonably ask four or five questions in thirty minutes. Even if you really want to ask more questions, be patient and let the conversation breathe. Sometimes detours can lead to interesting insights. Stay engaged and practice active listening. You can take notes, but ask politely if it is okay to do so. If you don't get through all your questions, you can politely ask about following up on email or scheduling another conversation. Always ask them to recommend a few other people in the field who can provide additional perspectives.

5. *Thank them.* Make sure to send a thank you note (via email or LinkedIn) within twenty-four to forty-eight hours after the conversation. Be specific and authentic; convey the highlights of the conversations. This is also a good time to ask any follow-up questions.

6. *Stay in touch and follow up.* A simple like or comment on the individual's LinkedIn post is better than nothing. Most importantly, please let people know if their advice, connection, or assistance leads to a successful outcome (e.g., a job interview or decision about pursuing a career). A simple gesture communicates the positive impact of your informational interview experience. People are more likely to accept other requests if they know their advice is helpful.

Typically, informational interviews in the United States are conducted between two individuals, one asking for information and the other offering it. This format presents a few barriers. First, the prospect of reaching out to individuals and conducting private conversations with strangers can be terrifying. Second, while international graduate students can rely on their US universities' alumni networks, foreign-trained postdocs in the United States may lack access to US alumni networks and therefore have less social capital. Third, people are so busy that professionals may not be able to entertain as many informational interviews as they want. Fourth, the individualistic nature of informational interviews can feel inauthentic to many international scientists.

COLLECTIVE LEARNING AND COMMUNITY-EMPOWERMENT

Community-centered knowledge sharing and mentorship has proven to be an effective and authentic model for international scientists who come from collective societies, as opposed to individualistic Western societies. Community-centered knowledge sharing is an age-old practice in many parts of the world. When I told my parents that I was interested in microbiology for undergraduate studies, my mother memorized the name of this new field. She kept bringing it up in her social circles until someone mentioned their niece was studying microbiology in Bangalore. She quickly got details and connected me with that individual. This was my first informational interview. That connection changed the course of my life as I was accepted in the microbiology program of the same college. In a community-centered model, groups of individuals collectively benefit from peer learning, mentorship, and communal forms of sponsorships. For example, consider a group informational interview where your lab hosts an alumni to learn together. In that case, anyone from your lab can contact the individual and set up a group informational interview at a convenient time for everyone. Make sure that the invited alum knows they will talk to a group.

Communal and collective learning models provide additional benefits, such as equity, fairness, and a culture of care, particularly for first-generation and international scientists. STEMPeers, a community-centered nonprofit serving eleven thousand members, is the world's largest international scientist network and an excellent case study of peer network–driven collective learning. STEMPeers' founders, Ananda Ghosh and Abhinav Dey, came to the United States for postdoc training after earning their PhDs in India. While exploring different careers, Ananda felt lost due to not having a professional network, credible resources for career development, and guidance from those who had knowledge of, or direct experience in, navigating the complicated US immigration system. They turned to other alumni from their PhD granting institution who were postdocs in the United States and realized the lack of resources and support was a widespread issue. To bring more visibility to the issue, in 2015, Ananda created the PhD Career Support Group (PhD CSG), a Facebook group, to convene international scientists. Abhinav was one of the early members of the PhD CSG and joined forces with Ananda to solve this issue through innovation. At first, the CSG primarily discussed the challenges surrounding immigration, problematic work environments for international scientists, and lack of social support. The platform embodied its name through collective career development and way finding. Ananda credited community-driven ingenuity for the growth and success of PhD CSG. He said, "You can build a robust, resourceful, and empowered community if it catalyzes collective brainstorming among people who are particularly motivated to get out of a bad situation." PhD CSG led to the formation of STEMPeers.

The nonprofit STEMPeers was born out of the need to find scalable solutions for the structural challenges international scientists face such as inequitable access to internships due to the immigration system and lack of career mentorships and role models. Such barriers mean fewer opportunities for international scientists to build a diverse portfolio of skills, experiences, and connections and access to mentorship. Collective efforts toward community empowerment resulted in STEMPeers initiatives such as Gurukool, Club SciWri, CSG Consultants, and STEMPeers conferences. Gurukool (a Sanskrit word meaning "place of learning") is a career mentorship program by international scientists for international scientists. Club SciWri is a science communication online magazine that started as a blog, where international scientists can build skills in communicating science to nontechnical audiences. The blog also increases the visibility of career stories, voices, and lived experiences of international scientists, including successes and challenges. By demonstrating their versatile communication skills through Club SciWri, a few international scientists secured jobs in medical and journal editing. CSG Consultants connects scientists interested in consulting and business development careers to pro bono consulting projects across borders, which later transformed into MedNess, aimed at providing current news and insights from the biotech world. All these innovative initiatives serve as a proxy for paid internships, which are largely inaccessible to international scientists.

To foster in-person interactions and learning among the community, they launched the annual STEMPeers conference in 2017. The STEMPeers conference also provided a network and a community to international scientists located in remote areas of the United States. The success of the US STEMPeers conference led to Euro STEMPeers, its European conference chapter. While inspired by the US STEMPeers conference, the Euro STEMPeers conference was curated to fit the cultural and professional contexts of Europe and STEM careers for internationals driven by UK and EU immigration policies.

As international scientists benefit from global mobility, these conferences foster global competition for high-skilled labor and workforce development. Science professionals in Europe learned about the US STEMPeers conference through the Euro STEMPeers conference. An employee of Spark BIH in Berlin learned about the US conference at the 2018 Euro STEMPeers conference. To connect with international scientists in the United States, Spark sponsored the employee to serve as a keynote speaker on careers in innovation and entrepreneurship in the next US STEMPeers conference. In addition to learning about career streams for scientists in the EU, US STEMPeers members also witnessed firsthand the crosstalk between international organizations.

These examples highlight the cross-continent benefits and impact of STEMPeers' collective empowerment model. First, it broadened the range of careers beyond research in academia and industry for international postdocs and graduate students, a breakthrough for the community. Second, this created opportunities for global workforce mobility, in turn instilling imagination, confidence, and, most importantly, agency in international scientists. Third, the appetite to create new paths and bodies of work inspired many new volunteers to come forward, enabling scaling of STEMPeers initiatives. The recognition, space, and innovative thinking facilitated the emergence of more international scientist-entrepreneurs. After 2018, STEMPeers started fund raising to expand access and opportunities for members through travel grants. As for the founders, Ananda found his way to the field of biotech equity research and Abhinav, to life science consulting and investment banking.

Stories like STEMPeers signify that international scientists can innovate by harnessing the model of community-centered knowledge sharing. Contrary to Western individualism, meaningful and successful international networks emerge from a rich tapestry of diverse lived experiences, a culture of care, and values of collective transformation and empowerment.

KEY HIGHLIGHTS

- Build a team of advisers, mentors, and sponsors with intention, based on your interests and goals.
- Use LinkedIn efficiently to expand your professional network and learn about fields.
- Informational interviews are an effective method of learning from someone's experience and knowledge.
- Communal knowledge sharing and collective learning models, centered on a culture of care, are authentic, inclusive, and equitable for international scientists.

CHAPTER SIX

Intentional and Nimble Professional Development

So far, I have outlined concepts and approaches for lifelong learning, a growth mindset, creative engineering, communicating strengths, and building networks for support and mentorship. You are likely thinking, "That's all well and good, but how can one reasonably apply these approaches on a busy schedule?" On the one hand, we are advised to strike a healthy work-life balance to avoid burnout; on the other, we are told to invest in lifelong professional development amid daily grinds. And on top of that, maintaining one's visa status and advancing immigration paths to permanent resident status and stability also takes time, effort, and money. All of this can feel overwhelming. I can attest that it's tricky to strike a balance between well-being and productivity.

I have earned the right to offer advice on this topic as a scientist who upskilled to change career fields and became successful, while navigating a meandering immigration path as an Indian citizen. I take pride in my ability to creatively design around restrictive systems and successfully implement approaches. The secret to my accomplishments is simple and not novel at all. I get a lot done by being strategic, intentional, and methodical and harnessing smart planning and consistent progress. In this chapter, I provide a systematic method for planning your future using an Individual Development Plan (IDP). Further, I highlight management approaches and tools to advance multiple priorities in a timely manner. Tactical approaches in this chapter will help you integrate concepts discussed in previous chapters.

As the realities of the immigration process may be closely tied to actualizing the goals of international scientists, professional development must also include strategic planning (short term and long term) to follow the immigration path. A typical strategy that many international scientists adopt is to focus on a practical professional path that enables moving from a temporary visa to permanent residency status (via the green card) first and entertain ambitious career goals once immigration isn't a barrier. While that is a sound strategy, it can also lead to enduring toxic work environments. Also, remember that you don't know what you don't know. That is why it is crucial to include an immigration plan in your professional development. Therefore, start early, and regularly assess and adjust your IDP.

WHAT IS AN INDIVIDUAL DEVELOPMENT PLAN?

An IDP is a process to outline and achieve your vision for the future by setting short-term and long-term goals and action plans by assessing your knowledge, skills, and interests; identifying areas of growth and development; and filling gaps to achieve success. More specifically, the IDP involves assessing one's capabilities, understanding expectations for professional success, and setting short- and long-term goals and action plans to bridge gaps and fulfill expectations. Research life and post-PhD careers are dynamic and unstructured. Therefore, you must take leadership in making progress, and IDPs can help with that.

Why an Individual Development Plan Is Important

The IDP is mandated by many training grants from the NSF and NIH and in turn is required for graduate and postdoc programs in STEM. Its fundamental purpose is to approach one's training or career goals strategically and intentionally through well-defined annual action plans and assessing progress annually. The benefits of an IDP include

- serving as a point of discussion and agreement on training and career goals between an adviser or supervisor and trainee;
- assisting in breaking down a complex and long-term future vision into short-term goals and concrete, actionable plans;
- clarifying roles and cultivating individual as well as group accountability;
- providing a structured framework in a dynamic environment, assisting in project and time management;
- providing transparency and benchmarking of progress via milestones; and
- having control over career planning and a positive impact on mental health.

As an international scientist, you must consider creating IDPs for your annual training or professional goals, as well as immigration and career goals. These can be separate documents but will often intersect.

How to Approach an Individual Development
Plan for Your Career

The process of crafting a career IDP can be broken down into the following steps.

1. Assessing your strengths, interests, skills, and values
2. Exploring and defining two or three careers of interest
3. Benchmarking success factors that are essential and desirable in your careers of interest
4. Inventorying your skills and experiences to identify gaps and areas for development
5. Setting short-term and long-term goals and action plans for development that will enable success

Online IDP tools for scientists such as myIDP by the AAAS and Chem IDP by the ACS facilitate creating dynamic IDPs. *Next Gen PhD* (Sinche 2016) includes helpful questions for self-assessment. Parallel to self-reflection, you should explore careers to identify options that align with your interests, skills, and values.

Exploring Careers and Defining Your Interests

There are many ways to learn about the landscape of careers for STEM PhDs and in your specific field. These include reading, listening to podcasts, or talking to people. Make a habit of attending alumni panels, career events hosted at your university or professional societies, and conducting informational interviews. PhD career seminars and symposia are becoming more common; many are organized online and open to broad audiences. A few notable examples are Princeton's GradFUTURES Forum, NIH's Career Symposium, What Can You Be with a PhD by a collection of universities in a tri-state area led by NYU, John Hopkins's PHUTURES, and UC Berkeley's Beyond Academia Conference.

As you would read new papers to expand your knowledge of your research field, regularly read job descriptions to learn about new career fields (Attig 2017). Do not worry yet about whether you have the required skills. Do this with the intent of understanding which kinds of jobs interest you. Job simulations, site visits, job shadowing, or experiential learning (e.g., internships, pro bono projects) also serve as good ways to explore careers. Depending on your bandwidth, you can select less time-consuming and flexible options such as job simulation exercises or more intensive ones like internships. Job simulations such as *InterSECT* are typically online interactive exercises that provide a taste of typical job duties in a range of science-related careers and a glimpse of the daily lives of PhD professionals (Nguyen 2018). I advise graduate students to consult early with advisers in their university's international office and graduate program administrators in their department to learn about ways to access internships. Equipped with know-how, you can advocate on your own behalf with your faculty adviser about participating in internships.

If career exploration feels daunting, find comfort in knowing that you can apply your research skills. Follow your curiosity, and write down questions. Also, treat career exploration like data collection and analysis, and be critical about information as you would about your research data. Be skeptical of extreme narratives. No job is all great or outright awful; jobs are generally a mix of both. Trust those who can provide a balanced perspective and then compare with your priorities. Like research, do not rely on single perspectives. Also, look for examples that share some of your specific contexts, for example, careers of other (visa-holder) international scientists.

Reflect on which fields align with your values and interests (Majumdar 2021). Self-reflection will assist you in eliminating jobs that aren't a good fit, allowing you to instead focus on those that bring joy and meaning. You should seek a balance between practical choices and professional fulfillment. Career exploration and defining your interest in two or three career fields is an iterative process and can take time, so start early. I typically advise graduate students and postdocs to do active career exploration at least one to two years before graduation. This gives you enough time for the next steps: benchmarking skills requirements, assessing your skills, and upskilling to fill gaps.

Benchmarking Success Factors in Careers of Interest

Whether reading job ads, talking to people, or engaging in experiential learning, you should benchmark required and desirable qualities for getting hired (and succeeding in the long term) in your top two or three careers of interest. Create a spreadsheet to inventory knowledge, skills, and acumen that are considered foundational, as well as those that are deemed distinguishing in each field. For example, when reading job ads, pay attention to the skills and qualifications that are required and those that are preferred. You can learn more from informational interviews. Ask which skills are essential and which additional skills and attributes will distinguish you from other candidates. Also inquire about baselines: what constitutes the basic, intermediate, and advanced levels of essential skills. This will clarify what you must have versus what is nice to have. After each conversation, make a note of it in your skills spreadsheet. You can give higher scores or weights to skills that are consistently desired for positions of interest by employers across sectors. Create a separate tab for each career field of interest and evaluate which skills are shared across fields and which are distinct. Skills benchmarking is important as shared skills will help you cast a wider net during your job search, and therefore you should prioritize building fluency in those skills.

Skills Assessment and Upskilling

Next, you must assess to what extent your scientific training has equipped you with what you will need in the careers of interest. You may want to gain additional skills and experience to match the level expected for success. Many scientists suffer from imposter syndrome, making them feel less qualified than they are. Trust your abilities; a PhD is a degree in self-learning. You also develop diverse skills beyond your domain knowledge during PhD or postdoc training. Use STAR storytelling to identify those skills. Consult with career mentors on which skills to develop. To upskill, look for free options first. Explore professional development programs, LinkedIn Learning, or Coursera courses that your university (or employer) provides free access to. Also, audit courses offered in other departments. For instance, if you are interested in science policy, consider auditing a policy course offered at your university. Finally, if your university provides opportunities for experiential learning such as internships, externships, project-based learning, or site visits for international students, take full advantage of them. These opportunities can be time-intensive but also carry high rewards in terms of gaining professional skills and networks. Create a tiered plan for upskilling based on effort and time required, where tier 1 is less intense (e.g., a one-hour seminar or site visit) and tier 3 is highly intensive (e.g., a three-month internship).

Finally, do not overextend yourself trying to upskill for all the qualifications for a specific career. Typically, employers invest in onboarding, mentoring, and professional development of prospective and current employees (e.g., Bridge to Consulting programs offered by consulting firms). You can garner employers' trust and appreciation by conveying your commitment to lifelong learning by mentioning prior experiences such as courses or internships during interviews.

The Art of Setting and Managing Goals

If you are feeling overwhelmed after reviewing the IDP process and wondering how to realistically achieve it on top of training or job duties, don't worry. Good management practices can help. The key is to manage yourself amid your circumstances. Set short-term (months) as well as long-term (years) goals and give yourself ample time by starting early. In the following sections, I outline how to move from complex goals and objectives to weekly tasks. In *The Practice of Management,* Peter Drucker popularized the concept of management by objectives (MBO), or implementing a strategic plan (Drucker 2006). Essentially a strategic plan involves outlining strategy, approaches, and tactics. Accordingly, every project plan is organized around a central goal, key objectives, and tasks (figure 5). Research projects are also organized in this manner. For example, your larger research question or hypothesis constitutes the central goal of your research project. This goal is broken down into two to three objectives (specific aims in your research plan), such that accomplishing any objective gets you measurably closer to realizing your goal. Each objective is further broken down into actionable tasks (e.g., experiments in your research plan). You can order the tasks under each objective based on dependencies between tasks. For

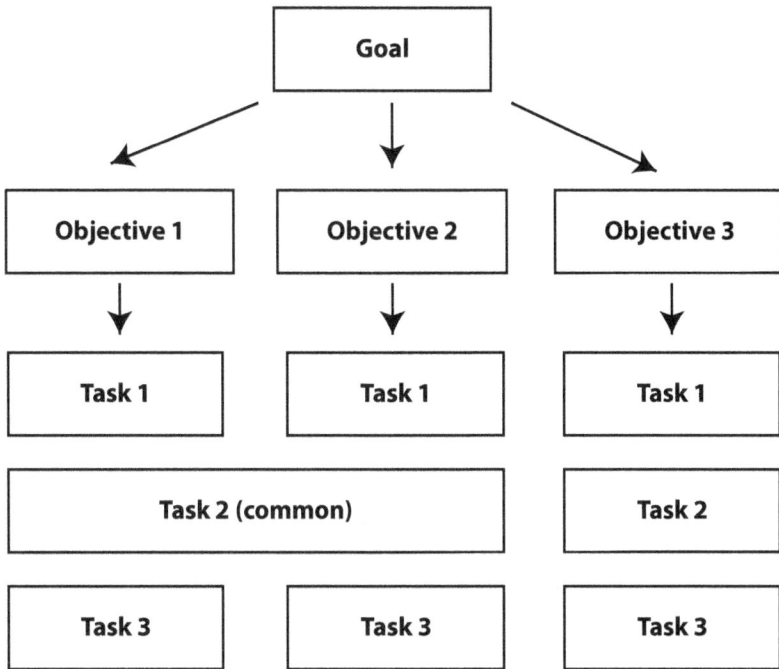

Figure 5. Schematic showing a typical organization of a project into goals, objectives, and tasks. Strategic planning involves identifying connections (or overlaps) between tasks.

example, does task 1 need to be done before or concurrently or after task 2? The most effective way to scaffold a project is creating an if-then logic enabling hierarchies or order of operations. If all tasks are completed, an objective is achieved. If all objectives are successful, you accomplish the goal.

All projects benefit from setting clear and effective goals. MBO provides frameworks for setting goals. One such framework is SMART goals:

Specific: Make your goal clear, concrete, and methodical.

Measurable: Create metrics and milestones to assess successful progress and completion.

Achievable: Assess whether your goal is realistic within capabilities and constraints.

Relevant: Is it important and relevant, now or in the future?

Timebound: Create a deadline and timeline for achieving your goal.

Example: SMART goal to explore consulting careers

I will learn about consulting careers for scientists by conducting one informational interview every week for three months and participating in events hosted by my university's consulting club. I will connect with PhD consultants in different kinds of consulting firms (from big to niche firms) and document highlights of interview conversations and assess whether it aligns with my interests and skills and my key development area. After three months, I will upskill by practicing cases with consulting club peers so I am ready for the hiring cycle in about nine months.

Assignment #1

Can you create a SMART goal for your next career step?

THE IDP IS AN ITERATIVE PROCESS

Creating an IDP is an iterative process whereby a downstream step (e.g., benchmarking) can inform an upstream step (e.g., defining career interest). For international scientists, each step of a career IDP will likely overlay specific contexts involving immigration and socioeconomic considerations. In addition to your professional skills and interests, self-assessment in step 1 will benefit from assessing specific strengths and values that your international upbringing and perspectives bring. For example, you could embed lessons from chapter 4 on identifying and communicating the specific strengths of international scientists. Step 3 of benchmarking success factors for careers of interest will also involve understanding immigration pathways to enter and advance in those career fields (see details in chapter 7). More specifically, learn about the kinds of visas and credentials that are required for successful petitions. Similarly, inventorying your professional portfolio (step 4) to identify gaps will involve outlining your immigration portfolio and identifying gaps to achieve short-term and long-term immigration paths. Finally, goal setting and action planning in step 5 will require embedding steps to strengthen your immigration portfolio and circumstances for career success in the United States or beyond.

By now I hope you appreciate that you need to set clear goals with milestones and deadlines for career development. You also need to allocate time each week to consistently make progress on your goals. With careful planning and consistent allocation of time, you can accomplish more than you give yourself credit for. As scientists, you manage research projects. Consider your career development and immigration goals as projects too. There are many online task management tools such as Asana and Trello that help you create a project plan with objectives, tasks, and milestones. If you are not entirely confident about your time management skills, I illustrate how to manage time in the next section.

Time Management Tips to Manage Diverse Goals

Whether it's New Year resolutions or career goals, people don't follow through either due to poorly outlined goals or time management issues. While urgency is a good driver, it does not help for career goals as those are incremental, iterative, and may require months or years. The silver lining of navigating the convoluted US immigration system is that internationals acquire practice in managing short- and long-term goals. We are forced into making short-term incremental plans toward long-term immigration goals. Like strategic planning, time management is a learned skill. With effective time management, you can get a lot done in your active hours.

PRIORITIZATION   The Eisenhower matrix (figure 6) is a powerful method for prioritizing tasks. Essentially, you classify tasks on two parameters: urgency and importance. Urgency is of course driven by

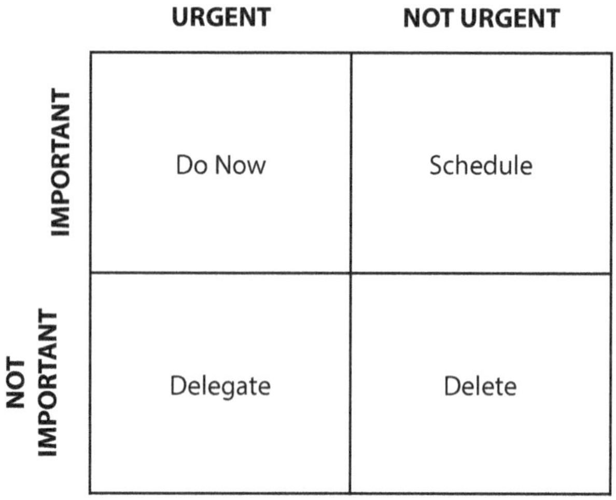

	URGENT	**NOT URGENT**
IMPORTANT	Do Now	Schedule
NOT IMPORTANT	Delegate	Delete

Figure 6. Eisenhower matrix of task prioritization.

deadlines. Do the tasks that are both urgent and important within the week. For tasks that are important but not (yet) urgent, schedule time in your calendar so you make incremental progress. If you do not, those tasks will move to your urgent and important quadrant very soon. For tasks that are not important (for you, your loved ones, or decision makers) but may be urgent, you may want to explore whether you can delegate or collaborate or get help. And for tasks that are neither urgent nor important, delete them from your plan. I need to add that time spent on things that bring you joy and physical and mental well-being is important.

SCHEDULING   I live and breathe by my work calendar. Once a week, I review immediate (weekly) and medium-term plans (monthly). For the important but not urgent category, I review my plans and schedule tasks in my calendar. Reverse engineering action plans will help you create deadlines for each step such that you make consistent progress and meet those deadlines. When scheduling, always add buffer time. If you think something will take one hour, budget seventy-five to ninety minutes. If something goes wrong, you have time to adjust and switch gears before the next task. I also color code my tasks so I know which category or project they fall in.

Identify times of the day when you are most focused and intellectually engaged. Schedule the most challenging or complex task at that time. For example, don't waste your productive time on answering emails or surfing the internet.

OVERCOMING PROCRASTINATION   Finally, if you tend to procrastinate, there is a larger underlying issue. Examine the cause: Are you overwhelmed or bored? Do you have a mental block? Your approach to managing procrastination will vary accordingly (Bailey 2017). For overwhelming projects, break them down to manageable steps. Get started; that's one step toward progress. There's truth to the adage, "Perfectionism is the enemy of progress." Cultivate comfort with imperfect versions. Also, seek help from mentors, and remember to exercise the growth mindset. If you get distracted easily, try the Pomodoro technique of short bursts of focused work followed by regular breaks (Collins 2021). Also, find an accountability buddy who will check on you. Finally, create rewards for completing milestones. Positive reinforcements lead to progress and outcomes.

As I mentioned earlier, dynamic task management tools such as Asana and Trello are very helpful for time management. You can assign start dates and deadlines for each task and sync it with your calendar. You can also delegate or add collaborators on tasks. In addition to automating the process, these tools keep you accountable and on pace for completing tasks. They also make it easier to iterate and adjust plans when things change. Finally, with the help of AI tools, you can speed up some tasks and gain time.

Decision Making

Strategizing and planning often require making tough decisions due to inherent risks and unpredictability that are outside your control. Whether it's research, career, or immigration, you can plan for what you can control and manage risks for what you cannot. Even for things you can control, strategic planning requires regular assessment and decisions such as which tasks to include, delete, and prioritize. Determining return on investment (ROI) is helpful when deciding on tasks. Your "investment" can take the form of effort, time, or money. As scientists, particularly graduate students and postdocs, we aren't financially secure. So I suggest not spending money when reasonable free or cheaper options are available. Time is scarce too, so select activities that give the most benefits or returns for your effort and time. For example, if you have decided on a specific career field, a paid internship, although time-intensive, may provide the most ROI due to the deep experience and access to professional networks it provides.

While making decisions, you'll also make trade-offs, compromises based on your circumstances. A typical example of trade-offs for internationals is accepting a job with lower compensation because the employer agreed to sponsor an employment-based green card rather than an ideal job with higher compensation that only provides short-term visa sponsorship. Immigrant life is a collection of such trade-offs, so we have practice. But don't let those trade-offs limit your options.

My final advice is to invest 10% of your weekly work time on your professional development goals. I spend three to four hours (mostly over the weekend) on professional development. In the recent past, it involved actively planning and advancing my immigration portfolio. More recently, my professional development time has included writing this book. It was an almost two-year project, but I kept my eyes on each month's goal and consistently spent time writing (three of four weekends). I reserved one weekend each month for relaxation and social activities.

Slowly and steadily, I am getting better at work-life balance. I sleep in on weekends and spend time on self-care and things that bring me joy and peace. I am also making incremental progress on not feeling guilty about enjoying life and not being on an endless productivity streak. It is a work in progress, a habit from PhD life that's hard to break. But I firmly believe that idleness fuels creativity and imagination. Some of my best ideas come on work-free weekends.

So go ahead, take that next step and create your IDP.

Assignment #2: IDP for Parallel Lives—Career Goals In and Outside
the United States
Map parallel futures by creating IDPs that may converge or diverge from each other, following these steps.

1. Storyboard parallel futures and aspirations (e.g., in the United States vs. your home country vs. a third country).
2. Gather requirements and information to inform goal setting.
3. Set short-term and long-term goals that balance the ideal and practical for each path.
4. Assess strengths and areas of growth to create action plans for each path. Action plans can inform a narrow focus on specific short-term options while maintaining the agility of pivoting in the long-term.

KEY HIGHLIGHTS

- Use Individual Development Plans to be intentional about your training and career goals.
- An IDP for career development is an iterative process involving self-assessment, career exploration, benchmarking success, assessing skills, and setting professional development goals.
- Set SMART (Specific, Measurable, Achievable, Relevant, and Timebound) goals.
- Break down goals into key objectives and then break down objectives into tasks. Use if-then logic to organize goals, objectives, and tasks. Include milestones to measure progress.
- Tips for effective time management include prioritization, scheduling, and methods for reducing procrastination.
- When choosing priorities, consider return on investment: time, money, and effort.

Sum of Choices, Not Constraints

CHAPTER SEVEN

Employment-Based Immigration Pathways in the United States

Like many international scientists, I moved from one visa to another through my seventeen years in the United States. My official label was "nonimmigrant," a visitor (a marked improvement from the previous "alien" label), until I became a permanent resident. I outline the different immigration options to study, train, work, and live in the United States and, someday, truly belong.

There are multiple visa types that are available for employment-based immigration pathways (figure 7). I must clarify that I am not giving legal advice but rather using public data to outline the immigration paths available to international scientists. For simplicity, I have classified the commonly used temporary visas in broad categories by purpose: training and work.

Optional Practical Training

If you are enrolled in a doctoral program in the United States under a student (F-1) visa, you can use Optional Practical Training (OPT) for employment or further training for one year before or after completing your PhD (USCIS n.d.g). If you are employed (part-time or full-time) before receiving the degree, you will use the pre-completion OPT for up

Temporary visas

Permanent Residency

Training	Work	Green card

- F-1 OPT (STEM Ext.)
- J-1 visa

- H-1B (cap)
- H-1B (cap exempt)
- O-1

- EB-1 (A,B,C)
- EB-2 (PERM/NIW)
- EB-3

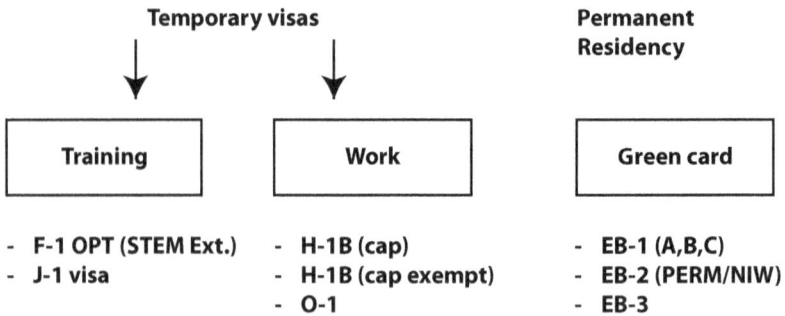

Figure 7. Immigration options for training and employment in the United States.

to one year. If you seek employment after receiving the degree, you will use the post-completion OPT for one year. The one-year OPT clock is fixed, which means that time spent on a pre-completion OPT will be subtracted from the one year toward the subsequent post-completion OPT. Therefore, be intentional about using the pre-completion OPT. For internships during PhD training, you should use the Curricular Practical Training, or CPT, instead of the pre-completion OPT (USCIS n.d.c). STEM PhD holders can also apply for a twenty-four-month extension of OPT, under the condition that their employers participate in an E-Verify program. The eligible programs are listed on the Department of Homeland Security website (USCIS n.d.a).

You can apply for OPT through your university's designated school official. Typically, international students and scholars offices or graduate schools handle OPT applications for F-1 visa-holding students. You can apply for OPT up to ninety days before or up to sixty days after your date of degree completion (or I-20 end date). Once your OPT application is approved, you will receive an Employment Authorization Document (EAD). The timing of applying for OPT is tricky, as you must plan for it around the same time that you are looking for a job. If you apply for OPT without a job offer, you need to secure a job within sixty days from the start date on your EAD. The speed of processing complicates matters further, as processing times may exceed ninety days in worst-case scenarios. Therefore, a slow processing time may delay your start date, particularly if you apply for OPT after securing a job offer (in the sixty-day window after graduation). This complex and unpredictable application timeline adds stress to an already stressful time of dissertation completion and job search. Therefore, you should consult with your international student adviser on when to apply for OPT. That said, even with the best planning, be prepared to pivot. In my case, I applied on the first eligible date (ninety days before my PhD completion). Even so, my EAD was delayed. As a result, I had to move my postdoc start date by a week. Therefore, you need to take charge of the process by regularly checking in with your university's international office, as well as clearly communicate such scenarios with your future employer.

J-1 Visa

J-1 (exchange visitor) visas are another category of temporary nonimmigrant visas typically used for teaching, research, and training purposes (USCIS n.d.b). The US Department of State (DOS) designates public and private organizations to act as sponsors for the Exchange Visitor Program. The majority of foreign postdoctoral researchers use J-1 visas, especially if they did not receive PhDs in the United States and cannot use OPT. International scientists who received PhD in the United States can also use J-1 visas after OPT. While primarily used for training, J-1 visas can also be used for temporary work, especially research. In 2022, DOS announced the Early Career STEM Research Initiative to increase the flow of talent in STEM fields by connecting certain J-1 program sponsors with STEM-relevant sponsors, including businesses (DOS n.d.b). While the length of J-1 visas varies by type, those for research (e.g., postdocs, visiting researchers, professors) are issued for a maximum of five years. In some cases, your sponsor can apply to DOS for an extension of the J-1 program (DOS n.d.a, n.d.c).

After the J-1 visa expires, researchers have a thirty-day travel period, commonly referred to as the "grace period," to depart. Depending on your country of origin, you may be required to return to your home country for two years at the end of the exchange visitor program. There is an option to apply for a waiver to the home residency requirement, but the process can be long (DOS n.d.d). You must apply for a waiver if you intend to continue working in the United States on a work visa. Therefore, plan early in your J-1 visa period, leaving enough time for the waiver process.

H-1B Visa

The H-1 visa is the most popular temporary, nonimmigrant work visa in the United States. Your prospective or current employer must petition for the H-1B on your behalf. Typically, employers of international scientists petition for a specialty occupation H-1B visa for theoretical and practical application of a body of highly specialized knowledge (USCIS n.d.e). The minimum eligibility requires a bachelor's or higher degree in the specialty occupation. To obtain an H-1B visa, your employer (petitioner) first applies for a Labor Condition Application (LCA) certified by the Department of Labor (DOL), followed by a Petition for a Nonimmigrant Worker with the US Citizenship and Immigration Services (USCIS). To make a case for a specialized occupation, the associated job description or position must fulfill the minimum required educational degree and include specific duties that are specialized and complex and align with your unique qualifications and expertise. An international scientist can enter (or continue employment in) the United States for three years while holding a H-1B visa. The H-1B visa can be extended for up to six years.

H-1Bs visas can be capped or cap-exempt. Basically, the cap implies an annual numerical limit of 65,000 new visas (each fiscal year) and an additional 20,000 visas for those with a master's degree or higher from a US university. The visas are selected through a lottery system. Every year, there are a few weeks in March when petitions for capped H-1Bs can be filed. The filed petitions selected through the lottery are invited for H-1B submission. As the number of applications each year are far more than available visas, you may not get it on the first try. Therefore, you (and your employer) should begin planning your path to H-1B early. If you are on OPT, you could start the H-1B process two years before your OPT expires. That way, you can apply for an H-1B a second time if not successful the first time. The other type of H-1B is called the cap-exempt H-1B or academic H-1B. H-1B petitions from universities or institutions of higher education or nonprofit and government research organizations are not subject to the numerical cap. They can be filed year-round rather than at a fixed time. This flexibility enables many international scientists to pursue education and research careers in the public sector. While review and processing of H-1B petitions can take months, your employer can expedite the process via premium processing, which includes paying additional fees. Petitions filed under premium processing are decided within fifteen business days.

The H-1B Modernization Rule, effective in January 2025, allows international entrepreneurs to apply (self-petition) for H-1B even if they own major or controlling interest in the company. This new rule is intended to facilitate international (foreign-born, visa-holding) entrepreneurs to start and grow their business if they fulfill key conditions (DHS 2024; Katz 2025).

O-1 Visa

The O-1 (or O-1A) nonimmigrant, temporary work visa is meant for individuals who possess extraordinary ability in the sciences, arts, education, business, or athletics (USCIS n.d.f). The O-1 visa is issued for three years and can be extended. As this is another employer-sponsored work visa, your employer will file an I-129 nonimmigrant petition on your behalf at least forty-five days before the start of employment. USCIS evaluates extraordinary ability in the sciences as a level of expertise achieved by a small percentage of professionals at the top of the field (USCIS 2023a). Supporting evidence to prove extraordinary ability can include the following:

- Nationally or internationally recognized prizes or awards for excellence in the field.
- Membership in associations in your field that require outstanding achievements of their members.
- Published material in professional or major trade publications or major media about your work in the field.
- Participation on a panel, or individually, as a judge of the work of others in the same or in an allied field of specialization.
- Evidence of original contributions and authorship in scientific, scholarly contributions of major significance in the field.
- Evidence of employment in a critical or essential capacity for an organization or of employment commanding a high salary

If these criteria seem like the ones for employment-based permanent residency (a green card), it is because they are, with subtle differences. The level of competency and strictness of criteria differ. That's why it is prudent to consult with an immigration attorney who can advise you on building a portfolio for different visa types or permanent residency.

SINGLE VERSUS DUAL INTENT TEMPORARY VISA   It is important to know the difference between temporary visas with respect to "intent" while planning your long-term immigration pathway. Single intent visas such as F-1 (student) and J-1 (exchange visitor) emphasize "nonimmigrant" intent. In other words, applicants for a single intent visa agree to use the visa only for the stated specific purpose (e.g., study or scholarly exchange) and leave the United States after completing the purpose. In contrast, dual intent visas (e.g., H-1B) allow holders to maintain their nonimmigrant status while also exploring a pathway to permanent residency in the long term. This distinction is important if you would like to stay in the United States for the long term. In that case, you may explore moving from a single intent to a dual intent visa before filing for permanent residency. Why is this important? When you enter (or reenter) the United States or go for visa interviews, you will be questioned about your long-term plans. Holders of single intent visas are expected to communicate their intent to return to their home country after a temporary stay in the United States Also, while you can technically file for permanent residency on a single intent visa, this is counter to the purpose of the visa and can make overseas travel (and US reentry) risky while your petition is pending.

Employment-Based Green Cards

To attain permanent residency, international scientists typically pursue employment-based (EB) green card pathways (USCIS 2022). As PhD holders, international scientists largely use two of the three EB "preference immigrants" categories: EB-1, first preference; or EB-2, second preference. The EB-1 category has three tracks. These are adjudicated on the following broad criteria:

- EB-1A: extraordinary ability in the sciences, arts, education, business, or athletics (can be self-sponsored)
- EB-1B: Outstanding professors and researchers
- EB-1C: Certain multinational managers and executives

The EB-2, or second preference, visa is given to nonimmigrant professionals who hold advanced degrees or who have exceptional ability. There are two ways to apply for an EB-2: request a National Interest Waiver (can be self-sponsored) or go through PERM (Program Electronic Review Management, administered by the DOL), also called Labor Certification. The EB-5 visa is another avenue for international entrepreneurs and investors. Those who have invested or are investing about $1,050,000 in a new commercial enterprise that will benefit the US economy and create at least ten full-time positions for qualifying employees are eligible for EB-5 green cards.

The entire permanent residency process is long and depending on preference category and your country of birth, can even take decades. For any employment-based tier, your employer (or you if self-sponsored) will file an I-140 form for immigrant petition for alien worker and justify your eligibility through a combination of the criteria. For the EB-2 via PERM, the I-140 is submitted to the USCIS after a successful Labor Certification process. Like H-1B, for PERM you must justify that you are uniquely qualified for a specific position compared to US citizens. Putting the case together can take many months, and processing times following submission can take over six months. Processing of my PERM petition by my former employer took about nine months after submission. The applicant receives a "priority date," which is the date of filing of the initial petition (PERM or I-140) that signifies your place in line. There is a premium-processing path for I-140 adjudication within fifteen business days on paying additional fees. Approval of I-140 indicates that you are eligible for permanent residency. This is the most competitive step, particularly in the EB-1 (first preference) category.

The final step involves application for a green card or Adjustment of Status using form I-485. Adjustment of Status is the process that you can use to apply for lawful permanent resident status if you are already present in the United States on a temporary visa. The I-485 (Adjustment of Status) step is the biggest bottleneck because of the backlog. US immigrant visas have country caps, which means there are annual limits on the number of green cards allotted by country, greatly disadvantaging applicants from China and India, the top two countries of origin of international scientists in the United States. To regulate the process, USCIS allows filing for an I-485 if the priority date for citizens of a specific country is "current" (i.e., before the listed date) in the monthly visa bulletin. For example, if the visa bulletin shows September 2020 as the filing date for Indians, you can file for an I-485 if your priority date is before September 2020.

At the current pace, international scientists from India who petitioned under EB-2 must wait over a decade to file Adjustment of Status.

In the ensuing years (or decades), those with an approved I-140 but wait-ing in the I-485 backlog can keep renewing their work visa (e.g., H-1B) every three years until they can get their physical green card. It may sound like a fair middle-ground solution, but it is not. H-1B holders must leave the United States to get their visas renewed at a US embassy or consulate abroad, severely restricting international travel, particularly if your H-1B expires and there aren't available appointment dates at the consulate or embassy. I suffered terrible consequences because of this. I couldn't travel home to attend my parents' last rites and funeral as I was stuck in the United States with an expired H-1B (but current work per-mit) and the earliest available appointment at a US consulate in India was four to six months later. Finally, in 2024, USCIS announced a pilot program enabling internationals residing in the United States to renew their visas domestically. At the time of writing, it was still being tested at a small scale, but it provides a glimmer of hope.

Changing jobs while waiting for Adjustment of Status can be hard. Your new employer must agree to sponsor an H-1B visa in the short term and continue or transfer your green card application in the long term. While the US government has enacted a clause stating that inter-national professionals can change jobs if their I-485 process is pending for at least six months, without losing their place in line, employers tend to be very risk averse. The burden is on the international candi-date to manage the nerves of hiring managers and advocate at every turn, armed with facts and legal know-how. Speaking from personal experience, it is exhausting. Every job change is an uphill march; it takes a year to recover from it. That's why many international scien-tists, primarily from China and India, don't change jobs regularly or, in some cases, miss out on promotions. This systemic inequity is a deter-rent to advancement for international scientists and a workforce and talent development issue for employers.

Children of international scientists caught in the green card backlog face uncertain futures (FWD.Us 2024). While they grow up in the United States as dependents of high-skilled international scientists, they no longer have immigration status if they turn twenty-one before their parents receive green cards. Then they must either apply for a visa or leave the United States. Imagine being told you will need to leave or apply for a student or work visa to continue living in the country where you lived your entire life. Recently, USCIS enacted the Child Status Protection Act "to protect certain noncitizen children from losing eligibility to obtain lawful permanent resident status based on an approved visa petition by providing a method to calculate the child's age that considers when an immigrant visa number becomes available" (USCIS 2023b).

Many international scientists can be sponsored for permanent residency via EB-1B (or EB-1A), which is much shorter. Before the pandemic, there was no backlog in EB-1 for Chinese or Indian citizens, but in 2024, there was a backlog of three to four years for Indians. EB-1-B is only for researchers and is therefore not an option for those who moved to nonresearch scientific fields like business of science, science communication. or science administration and management. This subset can explore EB-2, which takes much longer, and/or build a highly competitive portfolio to apply for EB-1A eventually. Under the circumstances, it is advisable for Indian scientists to apply for the EB-2 early, to secure a priority date. Subsequently, after building a competitive portfolio, you (or your employer) can apply for EB-1A or EB-1B and import your earlier EB-2 priority date. The earlier priority date increases the likelihood that your date will be current (or close enough) and enables applying for Adjustment of Status, thereby reducing the wait time. That's what I did. I used my EB-2 priority date (2020) to concurrently petition for EB-1A and Adjustment of Status in 2024.

If your head is spinning at this point, welcome to the immigration roller coaster. The immigration path for an international scientist is designed as a maze with a long trajectory. To build a competitive portfolio to obtain a green card, it takes years. For example, if we add up the years for an international scientist from India who came to the United States for a PhD it can take 20 years (or more) to secure permanent residency.

Let's break it down: 5 years on F-1 (student) visa (length of a typical STEM PhD) + 1–3 years of OPT + up to 6 years of H-1B (work) visa = 14 years and then green card backlog (can take 10 years). Some international scientists achieve early successes during their PhD or postdoc training and can apply for green cards via EB-1 right away, but that is a minority. The majority navigate a long, winding road to the professional freedom and stability afforded by a green card. This complex journey also exposes international scientists to workplace abuse, where toxic supervisors often take advantage of the visa precarity to overwork their international employees or trainees. Unfortunately, international scientists fall prey to the toxic underbelly of academic culture. I interviewed Harmit Malik, professor at the Fred Hutchinson Cancer Research Center. He mentioned that he hears the worst cases of abuse because of immigration status being directly tied to studentship or traineeship, where supervisors kept the passports of international scientists hostage (clearly illegal). He remarked, "Immigrant supervisors can be the worst at abusing their earned privilege and trying to hold immigration over trainees, constantly threatening them with termination and deportations." Many of these threats are not grounded in reality or can be stopped by administrators, but many international trainees don't know their rights and are afraid to seek help. While nonacademic fields are better, immigration is used as a hurdle across sectors. I have heard many stories of international scientists being passed over for promotions and advancement because visa systems are complicated. Recent layoffs in the federal workforce and tech sector have made matters worse for scientists requiring work sponsorship. Therefore, resilience is a requirement for international scientists, not a choice.

BUILDING AN IMMIGRATION PORTFOLIO TO MANAGE YOUR IMMIGRATION JOURNEY

Employers must justify hiring an international rather than a US citizen to USCIS. Each visa type has its own stipulations and factors to measure and assess professional excellence required for sponsorship. Taking charge of your career also involves self-advocacy to employers. The first hurdle typically is convincing the prospective employer to petition for an immigration sponsorship on your behalf. Often, hiring managers want to hire international scientists but need to negotiate with HR and immigration offices. You can collaborate with the hiring manager to make a case for yourself. This is where having an immigration portfolio helps. Like a CV or teaching portfolio, an immigration portfolio frames your experience and accomplishments in language and metrics that aligns with the requirements of USCIS (Cépeda and Chernets 2022).

How to Build an Immigration Portfolio

Let's consider a scenario where your employer plans to petition a permanent residency under employer-based sponsorship tier 1B (EB-1B). The USCIS describes EB-1B for outstanding researchers or professors as follows: "You must demonstrate international recognition for your outstanding achievements in a particular academic field. You must have at least 3 years of experience in teaching or research in that academic area. You must be entering the United States in order to pursue tenure or tenure track teaching or a comparable research position at a university, institution of higher education, or private employer" (USCIS 2022).

USCIS uses the following criteria as evidence of being an "outstanding researcher or professor," and you must meet at least two of the six following criteria for a successful petition.

- Receipt of major prizes or awards for outstanding achievement.
- Membership in associations that require their members to demonstrate outstanding achievement.
- Published material in professional publications written by others about the noncitizen's work in the academic field.
- Participation, either on a panel or individually, as a judge of the work of others in the same or an allied academic field.

- Original scientific or scholarly research contributions in the field.
- Authorship of scholarly books or articles (in scholarly journals with international circulation) in the academic field.

You need to approach your training and work intentionally, framing your accomplishments according to these criteria. Therefore, your training and career IDP should intersect, include, or enhance activities for your immigration portfolio. For the EB-1B example, you could outline plans and milestones for the following as potential evidence.

1. Number of scientific publications in reputable peer-reviewed journals and citations.
2. Patents filed (or awarded) and fellowships or grants awarded.
3. Mention of your research in the press, media, or cover art.
4. Number of peer reviews of others' papers or number of times you have judged posters.
5. Invited presentations at national conferences.
6. Science communication (written articles, science outreach, podcasts, etc.) with quantitative metrics such as number of readers, viewers, or downloads.

To accomplish at least two of the six criteria above, you need to first select what best fits your circumstances and set SMART goals. Suppose you select a number of original scientific publications as evidence of original scientific research and peer review as evidence of participation as a judge of the work of others as your two achievable criteria. Then you should evaluate the parts of these criteria that are integral to your training and parts you will need to develop. For example, publishing your original scientific work in reputable peer reviewed journals is an integral part of your research. However, as visibility, not simply merit, drives the impact of one's work, you can take extra measures to increase the impact metrics such as citations (Fagbule 2018). You can increase the visibility and readership of your publication by sharing on social media, presenting at conferences, soliciting press or media highlights, or creating visual abstracts to communicate topics for researchers in related fields. Similarly, you should approach editors of journals to serve as an independent peer reviewer and ask your faculty adviser to include your name on peer review collaborations. Ask near-peers in international scientist communities like STEMPeers about which editors or journals in your fields encourage peer review by early-career researchers. Start peer reviewing early and often to build your portfolio. To create a diverse portfolio of judging other's work, you can serve as a judge for STEM posters in local K–12 schools as part of science outreach or serve on conference organizing committees of your professional society that includes reviewing abstracts to select invited talks and posters.

When I suggest such approaches to international graduate students and postdocs, many report lack of time due to lab work because they do not feel their advisers support their professional development. I often hear that advisers expect trainees to do essential professional development to advance immigration paths after work or on weekends. In chapter 12, addressed to advisers and managers, I talk about creating an explicit culture of care and support. That said, we internationals must take ownership of our careers because no one else will. Taking charge of your future involves planning and setting long-term goals for immigration pathways too.

STRATEGIC PLANNING   In chapter 6, I discussed strategic planning methods: organizing projects into SMART goals, objectives, and tasks. Let's look at how to apply SMART goals to build your immigration portfolio. As an example, consider that getting a successful EB-1B is your goal. Two or three of the EB-1B criteria outlined by USCIS are your objectives. Let's take one criterion or objective, evidence of published material in professional publications written by others about the noncitizen's work in the academic field. You can break it down into tasks.

"I will provide evidence of published material written by others by submitting

1. two media highlights of my scientific publication (title) in local or university news by (target date)
2. my professional profile featured in the career sections of two professional society or journals by (target date)."

MAPPING CONNECTIONS BETWEEN OBJECTIVES   While devising action plans for your immigration SMART goals, do not treat the objectives for immigration goals in isolation. You should approach them through a systems lens; assess the relationship or connection between key tasks that traverse or connect parallel objectives. Let's say that you consider each criterion that USCIS sets for EB-1B as a separate objective. In that case, you must assess the dependency between these criteria when planning your immigration portfolio. For example, media highlights of your work written by others is evidence of published materials about you by others. But media highlights are typically related to your scientific publication, a separate criterion on original work. Therefore, reverse engineer your action plan for each EB-1B criterion (separate objective) such that media highlights and social media impact metrics closely follow your recent scientific publication. Create an action plan with timelines for, say, which media folks you should connect with (and when) to discuss your forthcoming publication. When should you collaborate on drafting a media release? That way, media highlights and social media amplifications will closely follow your recent scientific publication or patent or award.

Assignment:
Create a strategic plan for your immigration portfolio.
Here are a few tips to get started.

· Craft a SMART goal that includes key objectives based on USCIS criteria.
· Prioritize two to three key objectives.

- Consider the following tasks that will fulfill any of the key objectives:

  - Learn how to effectively peer review for a certain journal via webinar.
  - Ask to collaborate on a peer review with your adviser.
  - Benchmark journals and identify editors.
  - Enlist as a peer reviewer in a certain open-access journal or email editor.
  - Identify conferences or events where you can judge posters, and schedule on your calendar.
  - Schedule time in your calendar to complete one independent peer review every quarter or semester.

From these broad tasks, you can create stepwise actions, for example, learn how to and practice peer reviewing, enlist as a peer reviewer, conduct an independent peer review or other form of judging. Then you can establish timelines for each task. Once you have an outline, you can create milestones that will inform progress of your goal (e.g., one peer review equals 25% progress). If you give this goal twelve months rather than three, you can plan well in advance and will not feel overwhelmed even if you were investing time outside of your workday.

SHORT-TERM VERSUS LONG-TERM IMMIGRATION PLAN   A common consideration is whether you can get a green card faster in an academic or nonacademic career. The answer isn't simple, but some of the criteria for EB-1A or EB-1B may align better with academic research. For scientists, academic research provides more opportunities for original scientific contributions, a key criterion. You can also publish in industry research, but those papers can include many authors. It is easier to prove that you have made an important contribution to a field by providing evidence of scientific publications, particularly if you are the first, second, or corresponding author. On the other hand, more patents are filed in industry than academia. But if you do applied research in academia, you can file patents too. There are an increasing number of research commercialization activities in academia. Academic employers have more flexibility to sponsor researchers using H-1B, J-1 in the short term, and EB-1 green card in the long term. For Chinese and Indian citizens, EB-1 is a faster path to obtaining a green card, due to lower wait times in EB-2. Therefore, there is a case to be made for persisting in academic research to build a robust immigration portfolio. Once you get a green card, your options open up. If you prefer to move into an industry role, particularly in research, for financial and work-life balance reasons, you must negotiate a path to long-term sponsorship at your place of employment sooner than later.

Once you stop doing research, things get complicated as you also can't access the EB-1B option primarily for researchers. I learned that the hard way. I had not considered that I would be ineligible for EB-1B in a higher education administration role, so I pursued EB-2 with a decade-long wait for Indians. Once I got approval for my EB-2 petition with pending Adjustment of Status, I kept building my professional and immigration portfolio. Because of my leading role in building new professional development programs and partnerships at two universities and scholarship and leadership in PhD professional development at the national level, I was able to build a competitive case for EB-1A sponsorship. My petition was approved in less than a week, and I got a green card six months later. My wait time for a green card decreased from ten years (in 2020) to six months (in 2024).

Ultimately, there is no formula. You can creatively build paths to permanent residency and immigration stability. It is crucial to consult with immigration attorneys to craft short-term and long-term plans. As with your career, plan early. Have one or two primary strategies, but be open to serendipity.

KEY HIGHLIGHTS

- Explore employment-based immigration options in the United States and build an immigration plan and portfolio with the help of immigration experts (attorneys).
- Optional Practical Training (OPT) up to three years is available for US STEM PhDs.
- The J-1 visa is another research, training and exchange visitor single intent visa.
- The H-1B is the common work visa in the public (cap-exempt) and private sectors (capped).
- Scientists can obtain employment-based (EB) green cards in tiers 1 (EB-1) and 2 (EB-2) through their employers or self-sponsorship. Chinese and Indian citizens endure long waits, particularly in the EB-2 category, but can upgrade to EB-1 after securing a priority date with EB-2.

CHAPTER EIGHT

Job Search in the United States

I joke that I have a good track record piling up job offers because my visa situation compelled me to play the numbers game. Assuming about 50% of employers would sponsor immigration after making an offer, I made sure I had four job offers when moving from scientific research to career and professional development. In the past six years, I have hired and been part of hiring committees and learned the behind-the-scenes process of talent recruitment. In this chapter, I clarify the elements of the job search process and provide tips for effective applications, interviews, and negotiations, including negotiations for employment-based immigration sponsorship.

THE JOB SEARCH PROCESS
When to Apply

A question I get frequently from international graduate students is when to actively start the job search process. The answer is that it depends on two primary things: the career field and organization type and whether you have a dissertation defense date. First, most hiring and recruitment cycles are year-round and, with a few exceptions, are typically time-sensitive. Faculty jobs have a cyclical timeline when open positions are posted: summer to early fall, with applications due around October to December. The interviews are scheduled in spring, with offers made in late spring or early summer, followed by negotiation over the summer. Start dates for faculty positions can be a year after offers are made. Similarly, management consulting jobs in big firms follow a recruitment cycle, with interviews in fall. Like faculty jobs, start dates can be a year later. Outside of jobs that can have distant start dates, you should only start applying if you have discussed an estimated graduation date with your adviser and/or have started scheduling your dissertation defense date. Typically, applying for entry-level positions three months before your degree completion is a safe bet, especially if you have been exploring career paths for the previous six to twelve months. In that case, you have been building a professional network, skills, and experience and have a good idea about the kinds of jobs and organizations you are aiming for. Why is having an estimated degree completion date important while actively applying? Take the perspective of an employer. During the screening interview, recruiters will inquire about your degree completion date. You want to appear thoughtful and intentional, not just for their sake, but for yours too. Finally, cohort-based postdoctoral programs (e.g., science policy fellowships) have a defined start date, and eligibility requirements typically include degree completion by a specific date. As an international, you have a built-in benchmark for this three-month or ninety-day period, the earliest you can apply for OPT. Suffice it to say that you should actively apply for jobs right before or during your OPT application timeline.

Where to Find Jobs

If you are pursuing postdoc or faculty jobs, follow your scientific societies, sites like HigherEdJobs, and journals that regularly post those positions. Social media has made job searches across all sectors and fields more visible. Most jobs are posted on LinkedIn. You can search by title, field, location, education level, and organization. You can also look up contacts (e.g., PhD alumni from your university) who work in those organizations. In addition, you can let your professional network know that you are actively searching for jobs, and they may forward opportunities to you. There are career affinity groups (e.g., Future PI Slack) and LinkedIn groups (e.g., Strategy Consulting Network) that you should follow depending on your desired occupation. In addition, connecting with people working in your desired field or organization is important because they can tell you more about the organizational culture and give advice on your application materials, interviews, and negotiations. Some may even offer to refer your application internally or at least forward your CV or resume to the hiring units. Don't hesitate to ask, but don't feel offended if someone does not wish to do so. While applying cold for jobs is okay, your applications may not get past the Applicant Tracking System (ATS). The first hurdle is to make your application visible to the hiring manager past ATS and HR, and internal referral can help. Therefore, it is prudent to conduct informational interviews; you can contact the hiring team to learn more about the position before applying. They may tell you to apply first, but even so, they may look you up and keep an eye out for your application if they like your profile. Outside of faculty jobs, the most efficient job search process involves people, not cold emails. This is the singular hidden curriculum that international scientists need to learn. Instead of applying cold for fifty jobs, apply for twenty-five and use the remaining time to network with people who work in those or related departments in the organization. That said, please respect people's boundaries. They can give you information or advice and at best a referral, no more. Before actively launching into the job search, you should spend some time to understand general and specific recruitment and hiring practices in the fields of your interest via informational interviews. This habit will equip you with an effective job search and application strategy.

EFFECTIVE APPROACHES FOR JOB APPLICATIONS

It is imperative to take the employer's perspective during the job search. One of the fundamental mistakes job seekers make is centering themselves and reciting every ounce of information about what they have to offer. At every step of the process, whether application, interview, or negotiation, put yourself in the employers' shoes. Reflect on what they want. Do their priorities align with yours? Employers give hints about their priorities throughout, in job ads and during interviews. And you can find out more by simply asking.

Analyze Job Descriptions

Job ads tell a story. More importantly, they tell you what the hiring manager wants from the new hire. All ads outline features of the organization and positionality of the hiring unit or team within the organizational structure. You will find this information in the "About X Organization" and "Position Overview" sections. For example, "*The X role will report to XX in the office of XX. The office of XX is responsible for Y. The X role will execute, lead, design, Y₁Y₂.*" Using a systems lens is important for understanding the strategic priorities of the organization and the hiring unit and how the open position contributes to the goals at each level. In addition, the job duties and responsibilities also provide a window into the employers' priorities.

- What responsibilities do they highlight more or repeat?
- What skills or duties do they start with or emphasize or allot a higher percent of time to?
- Which skills are required versus desired or preferred?

Next time you read a job ad, highlight the key points, annotate the subtext, and perhaps speculate about what the employer is looking for. Make a note of these preferences, and tailor your application accordingly. Finally, if the job description gives a sense of the hiring team or unit, you could look up the current team composition if the information is public. The team composition will provide information about the combination of existing skills, qualifications, and strengths among the current members and gaps that the open position could fill. If you cannot find this information, you could schedule an informational interview with the hiring manager or recruiter to learn more about the role before applying.

TAILOR JOB APPLICATIONS

Once you have a broad idea of what the hiring team is looking for, you should tailor your resume (or CV) and cover letter to demonstrate how your unique combination of skills, experiences, and qualifications make you a good fit for the position. Remember, the goal of the application material is to secure an interview and perhaps serve as a conversation starter in interviews. Consider what you can include in your application to pique the interest of HR (or ATS) and the hiring manager. Because of the high volume of applications for competitive positions, the screening individuals may spend less than ten seconds on each application; AI screens faster. One rule of thumb is to include keywords and skills as they are listed in the job ad. That said, you need to communicate your unique combination of skills, experiences, and educational qualifications to stand out from the crowd and highlight your fitness for a particular role. Now that I have been on the hiring side, I can say it is easy to differentiate between a generic and tailored application. To stand out, tailor your application by weaving a narrative that responds to the employers' needs and priorities.

CV VERSUS RESUME   A CV is typically a history of your academic and professional accomplishments, whereas a resume is a much shorter summary of relevant skills and experiences (table 1).

Let me elaborate a few points that scientists typically struggle with.

PUBLICATIONS: While resumes are preferred in the nonacademic sector, you can submit a hybrid CV-resume (~2–3 pages) for industry research positions. For those positions, you can include a few relevant

TABLE I

CV versus Resume: Key Differences

CV	Resume
3–5 pages long, can be longer with years of experience	1–2 pages long
Commonly used in academic positions, grant applications, tenure, and promotion	Commonly used in nonacademic (for-profit or nonprofit) jobs
Lists all academic accomplishments (can include all publications, presentations, awards)	Primarily focuses on skills and relevant experience tailored to the position
Chronological history of education, research, teaching, and service	Outcomes oriented; nonchronological experience categorized by competency; can include professional summary

publications or presentations. For hybrid CVs, you can include a few publications and link to your Google Scholar or ORCID page for access to the full publication history. Building a personal website can be fruitful, as you can list all your publications, presentations, and other accomplishments in your preferred format and narrative. In that case, you can simply link your website in the header. For science-adjacent positions (e.g., business of science), outcomes and metrics such as number of collaborations, citation counts are preferred.

PROFESSIONAL SUMMARY: Many resumes include a short professional (or executive) summary consisting of one or two bullet points. Think of this section as the abstract of a scientific paper. How would you describe yourself in two lines that compel the reader to look at the rest of your resume? The professional summary can combine relevant technical expertise and human (soft) skills. For example:

· Biochemist with 10 years of experience investigating RNA-protein complexes.

- Leverage leadership skills to manage interdisciplinary teams and projects.

You can also write the professional summary in a paragraph format, but be careful not to exceed two sentences.

EXPERIENCE SECTION: You can organize professional experience chronologically by position titles or functionally, for example, research, teaching, leadership, communication, and so on. In either format, you can summarize each experience in two to three bullet points. Always start with action verbs (e.g., *managed*). Good bullet points tell a story, including goal, approach, and outcome.

Research Scientist | Institute | Duration

- Investigated composition and function of x systems involved in y.
- Developed x technique and computational pipeline to study y.
- Led an interdisciplinary research team resulting in widely cited (200 citations) manuscript in *Molecular Cell* (featured on cover)

INDUSTRY-SPECIFIC GUIDELINES: Some industries may have specific preferences such as the use of metrics-driven outcomes in experience, using specific types of outlines and templates. Professionals working in those fields can clarify these specifics. Many AI tools help you create resumes oriented to specific career sectors.

I keep a master CV that lists all my accomplishments and update it annually. From the master CV, you can curate relevant experience and skills to tailor your application for specific positions. Take time to carefully craft your resume and ask your career mentors to review and provide critical feedback. Finally, whether in a CV or a resume, please do not include personal information beyond your full name, email address, website or LinkedIn, and phone number. Even adding a mailing address is optional for resumes. Photos, country of origin, marital status, and so on, do not belong in a CV or resume.

COVER LETTER A cover letter is a form of persuasive writing to demonstrate why you are uniquely qualified and fit for the role. Essentially, it gives you a space for building a professional narrative and contextu-alizing your experiences by weaving components of the CV or resume. A cover letter also enables you to divulge your motivations, for exam-ple, why you did something in the past and how the open position con-nects to your future aspirations. Using simple language is an effective form of persuasive writing and storytelling. The primary goal is to make a compelling case for yourself by organizing themes and argu-ments, not confusing people by using fancy terms.

A cover letter is typically one page (two pages maximum). You should start with a short paragraph of one or two sentences summariz-ing your background and expertise and demonstrating interest in the specific open position (including the position title and ID). In the sec-ond (and third) paragraph(s), you should articulate why you are quali-fied for the role. You can highlight that you have one or two primary skills, qualifications, and strengths, particularly those required to suc-ceed in the role, by drawing on specific experiences listed in your CV as evidence. This is where you can also create a narrative about your motivations and elaborate on professional experiences beyond the bul-let points in the CV.

In the next paragraph, you should outline your vision for the role and what you might bring to it. You can also include areas of growth in the new position that would help the hiring team understand why you are interested in it. Candidates often lean so hard on explaining why they are qualified that they miss out on detailing how the position helps them grow professionally (beyond material benefits like salary). For instance, you might say, "As x (position title), I am particularly interested in building/developing y, which provides an opportunity to learn z, thereby assisting in my professional growth in an area of inter-est." The concluding paragraph, also one or two sentences, should affirm your interest in the position and welcome the opportunity to discuss it further (in other words, you hope you will be invited for an interview).

WHO TO ADDRESS THE COVER LETTER TO   Ideally, you should address the cover letter to someone specific instead of "To Whom It May Concern." If the job ad outlines which team the position will be part of or the position title of the supervisor or manager, you should identify that individual and address the cover letter to them specifically. If the ad does not identify the team or manager, there may be a section identifying the "HR or Talent Acquisition" professional handling the hiring. That section is typically worded, "Please contact X if you have further questions about the position." In that case, you could address the cover letter to the identified HR professional. It's a small gesture, but it shows that you took a little time to do your homework or read the job ad carefully and are serious about this position.

To make it look more professional, consider including your institution's letterhead. You are technically a part-time or full-time employee as a PhD student (research assistant) or postdoctoral researcher. The style and format (e.g., type font, font size) should match the CV as you will likely submit it as an application package. Finally, request a peer or mentor to review and provide feedback on your cover letter. I would advise including a cover letter in the body of the follow-up email to the hiring manager even if the job ad doesn't ask for it. Why wouldn't you want to grab the competitive advantage by providing more context about yourself?

INTERVIEW SKILLS

If you get invited for an interview, celebrate the moment. Congratulations! The hiring team likely believes you are qualified for the position, at least on paper. Through the interview process, they would like to make sure that you are as good in person and learn more about you to determine your fitness for the role, the hiring unit, and the institution at large. Therefore, during an interview, you must present a holistic picture of yourself, including technical expertise, human (soft) skills, and professional interests and values. We spend so much of our weekly lives at work that personalities and work styles matter in a team. Through interviewing, your potential colleagues would like to find a skilled, friendly, and self-aware person and a good departmental citizen. Self-knowledge helps when writing good applications and performing well in interviews; there is nothing more refreshing than an honest and self-assured candidate.

The duration of the hiring process can vary by institution or sector. Typically, interviews can be many rounds, including screening calls with HR and the hiring manager and Zoom and/or on-campus interview(s) with members of the hiring unit and multiple stakeholders. Interviews are typically a combination of behavioral and technical interviews. If the organization wants to hire quickly, employers may combine many of these in a one-day campus visit (or all-day Zoom interview) after an initial screening call.

The key to interviewing well is simple: prepare, prepare, and prepare some more. First, know the rules of the game by asking the following questions before every interview.

- How long is it?
- What is the format, via phone or video (Zoom) or in person?
- Who will you meet? Ask the identities of people you will meet so that you can look them up. For final rounds, which can be all day or several hours, ask for the detailed interview schedule.
- Who will organize logistics such as travel, overnight stay, and so on?

Be polite with individuals who organize the interviews. Remember that you are being assessed by everyone, no matter their job title, from the first email exchange. Therefore, no matter how busy or stressed you are, be kind and patient.

Before the Interview

You can prepare in the following ways.

RESEARCH   Research the field, institution, hiring unit, and people you will meet. Starting with the organization or institution, look up its mission, values, and goals. These are not just words; they inform strategic priorities. Look for any recent articles on or interviews of the CEO or chair or founders of the institution to learn about their future directions. In addition, read the latest news so you don't seem clueless if something big happened in the field. Next, look up the unit or team, including any public information about their work on the organization's website, in articles, or on LinkedIn. Look up individual members of the team and read their recent LinkedIn posts. Then carefully peruse the interview itinerary and look up every individual who will interview you. Try to understand the dynamics of each interviewer. If you don't understand why someone is included, make a note to ask questions about how they work with the hiring team during the interview.

ANSWERS TO TYPICAL INTERVIEW QUESTIONS   Every interview round may start with a common statement like "Tell me about yourself" or "Tell me why you are interested in this position." No matter how they phrase it, they want to know why you are interested in this specific position. Prepare a concise response that doesn't take more than two minutes. The response should include relevant past experiences that inform your interest. More specifically, and this is the hardest for scientists, consider carefully whether (or how much) you talk about your research. If you are interviewing for a position that is not related to research, your detailed research experience is immaterial. Choose honesty over trying to impress someone with your accomplishments. That said, focus more on what you would like to move toward than on what you are moving away from. This is particularly important if you are switching fields and moving out of research. Whatever the reason, focus more on this specific field, unless someone asks why you are leaving research. In my case, when I was moving from a postdoc to a PhD professional development field, I talked about how I got interested in PhD professional development and enhancing scientific training, how I explored the field, and the inflection points inspiring me to pursue it full-time. As I had worked before PhD training, I wove in my nonlinear career path to reflect the theme of my career and what motivated my career changes; for example, "I am driven by the desire to solve wicked (complex) problems, the nature of which may change." You need to think hard and communicate your career motivations and your interest in a specific role that aligns with your motivations. Prepare a concise and compelling response, and remember that you will have to repeat this response many times. Therefore, keep it simple and engaging enough that you don't get bored repeating it. You should sound just as enthusiastic saying it for the seventh time at 5 p.m. as you did the time at 9 a.m.

Other common interview questions include the following:

· What are some of your strengths?
· What do you anticipate as areas of growth or the learning curve in this role?
· How does this position align with your future aspirations? Where do you see yourself in five years?

For research-oriented roles, prepare a one-minute and a three-minute pitch of your research experience in accessible language, avoiding jargon.

BEHAVIORAL INTERVIEW The goal of behavioral interviews is to highlight your skills and strengths by demonstrating applied examples. Behavioral interviews typically start with these statements:

· Tell me of a time when you … (resolved a conflict/ managed multiple projects/ led multiple stakeholders), or
· How would you describe your (leadership style/ time management skills/ ability to collaborate or work in teams)

In chapter 4, I described the STAR method of skills storytelling and advised you to maintain a STAR journal of skills stories. If you follow that advice, you'll shine in behavioral interviews. On the day before interviews, brush up on the stories and think of which stories will land well for this job. You can also guess a few behavioral interview questions by looking at the job ad carefully, particularly in the desired qualities and skills section.

QUESTIONS ABOUT SALARY EXPECTATIONS   During any round of interviews, particularly with HR, you may get questions about your salary expectations. You can evade it if you like by saying that it depends on the whole compensation package, that you don't want to look at salary alone. If they insist, you should communicate the ideal range of salary. That's why before interviews, you should do some homework on the typical salary range for specific position titles in x field. Inquire about the salary range, so that you know what's a typical minimum, median, and maximum for the type of position (with desired years of experience). Knowing about the range will help you during negotiation. You can look up salaries on Glassdoor or inquire from mentors or people in your network familiar with the sector or field.

TECHNICAL INTERVIEW   Depending on the field, technical interviews can vary. For research positions, you may be asked to prepare a research talk in order to assess your research communication and data storytelling skills. For computational positions, your coding skills may be tested. CASE interviews are common in management consulting interviews. For science communication or science writing, you may need to write an article or create an infographic. For administrative and management positions, you may need to address a prompt about solving a specific problem in the field. In my field, we regularly set up professional development workshops. My technical interviews typically involved developing an outline of a workshop (including an overview of the curriculum and an assessment framework) or leading a mock workshop with students. You should prepare ahead for technical interviews too. For presentations, learn about the audience, and design your content based on how much time you have. You should consult with career mentors, alumni, or those in your network who have knowledge of the specific technical interviews in your field or organization of interest. Request them to help you prepare. For consulting CASE interviews, practice with peers and join your university's consulting club.

PREPARE QUESTIONS OF YOUR OWN   Interviewees are given some time to ask questions during every interview session. Use that opportunity well. Interviewers also assess the quality of questions you ask them, which conveys your genuine interest in and curiosity about the position and the employer's work. You can prepare a few questions that are common for everyone. Typically, these questions inquire about the culture of the teams and organization; for example, "What do you love about your job or this company?" Also prepare different questions tailored for specific individuals you meet, based on their roles. In addition to being assessed, you should also use this opportunity to learn more about the position, the people, and the organization to evaluate and determine whether this is a good fit for you. Therefore, use your questions well, to genuinely gather information to aid in your decision. When you are wrapping up your interview with the hiring manager or HR, remember to ask about the next steps and when you may expect to hear from them. This will help your nerves as you wait. If you don't hear back from them in the estimated time, wait another week and then follow up on the status of your application.

During the Interview

Stay calm and try to enjoy the process. Lead with genuine curiosity about the people and the position. That mindset will invariably make you more relaxed, and you'll perform well. Instead of thinking of the interview as a test, try to engage in it like a series of conversations. This brings me to the next tip: practice active listening and ask probing questions. Part of effective interview practices is listening well and engaging someone by asking good probing questions, not simply impressing with your response.

People will remember how you made them feel. So yes, vibes matter along with facts and expertise. If you don't know something, be honest. Don't waste their time by making something up; it's disingenuous and dangerous, particularly if folks interpret it as misguided confidence.

Please note that employers cannot lawfully ask you questions that lead to illegal discrimination on the basis of race, sex, age, religion, national origin, marital status, physical disability, or (in some states) sexual orientation. They will not ask personal questions. However, context is important. If someone wants to help you understand child-care benefits and asks you whether you have children, in that case they intend well. That said, you are always within your rights to say that you are not comfortable sharing details about your personal life.

Beware of the informal parts of the interviews, such as conversations over meals (or worse, alcohol) or walking with someone between interviews. These are tricky situations where you may feel comfortable and let your guard down. Remember, no matter how much fun the environment is, you are being assessed in these situations too. Even if others are being casual and engaging in controversial topics, don't contribute. You can be fun, but don't say anything you would regret later. Finally, all-day interviews are intense. You'll need to maintain your focus as well as your well-being in a stressful situation. Stay hydrated, nourished, and caffeinated (but don't overcaffeinate) to maintain energy and focus. In addition, take regular breaks to simply relax and breathe. Remember, you need to excel through the day, so pace yourself. In general, don't schedule more than two interviews in a week, as you will get exhausted and burn out. Pace yourself through the job search.

After the Interview

Send a thank you note (via email or LinkedIn) to at least the hiring manager and, if you like, every interviewer you meet. Do not make the note generic. Add something specific, perhaps an insight or discussion point specific to that person, but don't expect a reply.

Get into the habit of documenting key highlights of the discussions and your own reflections. The concept of organizational or team culture is nebulous, and people can project progressive values. Interviewers' responses to specific questions tell a lot about how they really function. Reflecting on such moments will confirm any red flags that you may have picked up subconsciously. On the other hand, you may truly enjoy the work if you had engaging discussions and great chemistry (it's kind of like dating) with your hiring manager or prospective team. I still remember my campus interview at UVA. I had a series of intellectually engaging conversations and discussions; I came out of it energized and knew I wanted that job. I had a great time in it.

If you don't get an offer, you can request feedback for improvement. I genuinely enjoyed meeting people during interviews without dwelling too much on the outcome. As a result, I interviewed well. Even if I didn't get an offer for some reason, interviewers liked me enough to recommend me to others. Remember, interviews are also an excellent networking opportunity. Whether or not you get the offer, you can always walk away with lessons, connections, or leads.

EFFECTIVE NEGOTIATION

Congratulations for getting the offer. Celebrate the moment before stressing about the negotiation process. Approach negotiation with a positive and healthy mindset. Your employer wants you and is on your side. Therefore, work with them to find a win-win situation. You can achieve a win-win by leading with an interest-based negotiation over position-based negotiation. What does that mean? A position is a stance that states what you want. An interest-based negotiation divulges the underlying reason or values behind the stance. Leading with a position would be demanding a specific salary. Focusing on interest would be conveying why you would like this range of salary, for example, cost of living in that city, childcare, health reasons. When your hiring manager knows your motivations and values clearly, they may help you by leveraging creative options.

Interest-based negotiations work because there are many items to negotiate beyond salary, which include

- Total compensation: salary and benefits (retirement, healthcare, employer match)
- Stock options (for private sector companies with IPO)
- Moving and relocation benefits
- Housing allowance
- Signing bonus
- Annual increase
- Vacation time
- Start date
- Childcare or dual-career support (spousal career support)
- Professional development allowance (e.g., conference travel)
- Long-term immigration support

When you negotiate in alignment with your values and priorities, you can get a good overall deal. Once you trust your hiring manager to be candid about your interests and values, also learn about what they can adjust and what they can't. Perhaps they can't adjust the salary much but may be able to increase your stock options or hiring bonus or relocation allowance. Money or value can come from different buckets. Therefore, be creative and collaborative to expand the pie.

Before you get into a negotiation conversation, you must be clear about your values, interests, and priorities. One way of achieving this clarity is to create a list of things under two categories, "must have" and "nice to have," and the reasons they are in those categories. Be honest with yourself. Also think long-term and determine the short-term trade-offs you are willing to make for long-term gains. A particular example for international scientists is the long-term immigration path. Will the employer support you beyond the OPT or H-1B visa? If they are willing to sponsor your green card, you could make peace with a slightly lower salary. Or, if you must have annual time off for one-month visits back home, adjust on other items. We often make these trade-offs.

For quantitative items such as salary, do due diligence. I mentioned before that you should benchmark typical salaries for the position or similar positions if the salary range isn't outlined in the job ad. A parallel approach for determining the lower limit of your salary range (also called the floor) is to do a cost of living (COL) calculation using online COL calculators such as NerdWallet. That amount is equal to a horizontal move (without advancement) in terms of salary and is your salary floor (not really, but the very minimum you can settle for). Your upper limit should be a little higher than your target salary; think of your target salary as the median. If they make an offer with a salary listed, negotiating 10% higher is reasonable. Based on your competitive advantage (years of experience, skills, and qualifications), you can negotiate further.

Once you have your must-have and nice-to-have list, you'll first negotiate to get your must-haves, followed by a few nice-to-haves. But if you fail to achieve your ideal outcome, create an alternative scenario that is another permutation of must-haves and nice-to-haves as the next best outcome. This is called a BATNA, which stands for Best Alternative to a Negotiated Agreement. BATNA is used when you can't reach a negotiated agreement on your desired outcome. In that case, what's your next best outcome? If you can't get the upper limit of the desired salary range, can you negotiate for a higher signing bonus to make up the difference that year? Or perhaps you can negotiate for a raise in six months to make up the difference.

HOW TO NEGOTIATE   When you get the offer, especially over the phone, hear the terms, but do not commit to anything. Do not try frantically to take notes; actively listen. Politely ask for a written version of the complete offer. If they ask about salary expectations, repeat that you would like to look at the total compensation package first. Once you get the written offer, you can ask for a few days to consider it before negotiating.. After you have evaluated and come up with a counteroffer or negotiation points, ideally set up a call. It is easier to negotiate verbally than via email. You can better convey politeness, tone, and appreciation. Always begin the negotiation with thanking them for the offer and expressing your genuine gratitude. You can then proceed to ask a few questions or express areas you would like to discuss. Negotiate items one by one. Gauge the person's tone and response to determine what they can or can't adjust. The hiring manager or HR may be able to make decisions on the spot or may ask for more time after hearing your counteroffer. Be patient; negotiations can take weeks or even months. Stay polite and professional throughout the process; remember that you will likely work with them. Once you get the final offer, you can ask for some time to decide. Typically, one to two weeks to think it over is reasonable.

If you are waiting to hear about the status of another interview, particularly if that is your first choice, you will want to let them know that you have an offer. Don't wait to follow up with other places. That way, the other employer has time to sort things out while you are going through the negotiation with employer 1. If employer 2 responds quickly, they are considering making you an offer. Depending on your preference, you may want to help speed things up with the employer of choice. If employer 2 asks you about details of your offer, share as much as you can comfortably. Typically, having another job offer provides more leverage and negotiation power. But don't abuse it. Remember that these are either your future colleagues or professional network in your field. You want to maintain a good professional reputation through the process. You can improve your negotiation skills with knowledge and experience. *Getting to Yes: Negotiating Agreements without Giving In* by Roger Fisher and William Uri is a universally recommended book on this topic (Fisher and Uri 1981).

NAVIGATING VISA SPONSORSHIP DURING THE JOB SEARCH

One of the hardest things to grapple with during the job search process is when to bring up the topic of visa sponsorship. It really depends;

great arguments can be made for stating requirements in the beginning as well as waiting to bring it up. One line of thinking is, why waste your and your employer's time going through the hiring process if the position isn't amenable to sponsorship? On the other hand, many international scientists bring it up after proving their worth to the hiring manager in the hope that they can negotiate on visa sponsorship. I have done both with mixed outcomes. Let's walk through each option.

Bringing up Visa Sponsorship Early

Typically, the HR application site includes a question such as, "Would you now or in the future need sponsorship to work in the US?" You should respond to it honestly. Even if you have three years of OPT sponsored by your graduate school, please answer yes. You may need sponsorship beyond the three years. One of the downsides of this question is facing automatic rejection if the position isn't set up for visa sponsorship. That is why it is important to connect independently with the recruiter or hiring manager. Inquire about sponsorship with them before applying. If they invite you to apply and the HR system auto-rejects, you can enlist the support of the hiring manager. They can communicate with their HR team to work around this issue. If there is no scope for sponsorship in the position, your hiring manager or recruiter will tell you. Let's say you responded "yes" to needing sponsorship during your application and your application moves forward. On a screening call, HR may ask you about sponsorship. If they do, be honest. If they don't bring up sponsorship, should you bring it up? Again, it depends. If you feel strongly that you do not want to waste your time and someone else's, bring it up early in the conversation. It's certainly easier to have this position when you have a stable job or have other offers. In my current professional stage, I inquire about sponsorship even before applying. I don't apply if they sound skeptical or they can't sponsor. That wasn't always the case, especially as I approached my career change after the postdoc. This brings me to the second option.

Discussing Visa Sponsorship in the Later Stages

In many jobs, you may not get clarity about sponsorship. The hiring manager may be unsure about sponsorship, particularly if they didn't discuss it explicitly with HR. In such cases, both sides take chances to see what is possible. You may focus on proving your value to the hiring team and once they are convinced, bring up sponsorship at the later stages, even during negotiation. I waited to discuss sponsorship when I was aggressively interviewing close to the end of my postdoc. I would do screening interviews and bring up sponsorship during campus interviews or final rounds. By then, I had the opportunity to prove my worth to everyone involved. Many times, hiring managers would say they will figure something out or discuss with HR when they want to hire you. In my case, some hiring managers were successful in making a case (with my help) for HR and the immigration office to sponsor a petition, but in some cases HR wouldn't budge. I mentioned before that I stacked up job offers as for me the litmus test was sponsorship approval from the employer. One of my offers was rescinded after the hiring manager tried hard but could not convince their HR team to sponsor my H-1B. It happens more often than you think; HR and hiring managers are not on the same page. You need to make peace with that risk if you invest in this option.

However, I succeeded in convincing two employers to make a compelling case for H-1B sponsorship by collaborating with hiring managers. You need to be prepared with facts to self-advocate. Show successful examples where other employers (particularly competitors) have sponsored. Consult with an immigration attorney to stay informed. Unpack the path to sponsorship for your hiring manager. Outline options with supporting links and information to your hiring manager, so they can negotiate on your behalf. I did market research on how many job ads in similar positions require or prefer PhDs and demographic analysis (e.g., percent of internationals vs. citizens with PhDs in specific fields). Once the employers agreed, they could build on my preparatory work for the petition to USCIS. My manager and the immigration officer at UVA spent months building my case. They would regularly reach out to convey challenges or ask for more data, and I would go hunting for facts and evidence. Patience is key, as these things can take time (sometimes months). I went through emotional ups and downs, including anger, resentment, frustration—the entire gamut. But the next day, I would put one foot forward to gather evidence and continue self-advocating. Remember, your determination instills confidence in your hiring manager too. They want you; they are on your side. Therefore, collaborate with them.

During negotiations, discuss short-term and long-term sponsorship paths to a green card. That way, your manager will know how to plan to retain and advance you. Another thing to watch out for is the feeling of guilt and embarrassment that can creep in as you see your manager work so hard to sponsor you. That feeling of guilt may make you hesitant about negotiating on salary or other items. But remember about pay equity. By not being paid what you are worth, you may unintentionally be setting a bad precedent for other international hires. Remember, you are doing a disservice, not just to yourself, but to the international community by conflating immigration support with fair compensation. Besides, you will have to pass the wage determination part to receive labor certification for your visas. We internationals contribute immensely to our workplaces and the economy and are worth every bit the effort.

KEY HIGHLIGHTS

- Regular informational interviews can clarify industry-specific standards of talent, recruitment, effective applications, and interview and negotiation strategies.
- Contact people in your professional networks for internal referrals.
- Take the perspective of an employer to tailor applications and interview strategies.
- Research the field, organization, department, and people to prepare for your interview. Ask questions to learn more about the expectations of the successful candidate and growth opportunity.
- Negotiation is not a zero-sum game. Communicate your interest, not position. Collaborate with your employer to find win-win solutions.
- There is no magic formula to navigating immigration sponsorships. Arm yourself with information to self-advocate. Also, increase your odds by diversifying your search.

CHAPTER NINE

Global Careers for Scientists

The United States is a global leader in recruiting and retaining the best scientific minds from around the world. However, in recent times, international scientists trained in the United States are increasingly exploring global careers because of the declining quality of life, combined with desirable incentives driven by the higher global demand for scientific talent. These incentives include easier immigration paths to permanent residency and a better socioeconomic climate. In this chapter, I highlight data-informed trends, global scientific workforce development programs, and opportunities for scientists in Asia, Australia, Canada, the European Union, and United Kingdom. I summarize approaches, resources, and considerations for exploring global careers. Through stories of international scientists who moved to other countries after PhD or postdoc training in the United States, I outline how to identify opportunities, do cost-benefit analyses, and approach migration in the short and long term for career advancement. Let's start with data, as we are scientists after all.

DATA ON GLOBAL MOBILITY OF SCIENTISTS

A 2021 statistical analysis of the global migration of 3.5 million scientists over sixty years reconstructed geographic career paths and provided a data-informed model (Vaccario, Verginer, and Schweitzer 2021). A distribution analysis of about 62,000 scientists who relocated between 2000 and 2008 indicated frequent relocations between cities that are more than 5,000 kilometers (3,016 mi.) apart. The study showed that relocation is frequent early in the career and decreases with "professional age."

Perception of Global Mobility among International PhDs

The Survey of Earned Doctorates (NCSES 2023, table 2-8) indicates recent PhD graduates' intention to continue their stay in the United States. In 2022, 74.3% of international PhD graduates (temporary visa holders) indicated intent to stay in the United States, up from 72.5% in 2016. When comparing international PhDs by fields, 16.8% of STEM PhDs reported career outcomes outside the United States, compared to 35.6% of arts, humanities, and social sciences PhDs (NCSES 2023, table 6-3). Thus, the United States still retains most international STEM PhDs it trains.

Next, I analyzed the responses of international (temporary visa holder) scientists in the United States representing the top ten countries of origin among the population (NCSES 2023, table 7-7). Fewer PhDs from China (down from 81% in 2016 to 76% in 2022) and South Korea expressed intent to stay in the United States. Likely reasons for this downturn include improving career prospects for STEM PhDs in China, South Korea and other parts of Asia due to economic growth and investments in research and development (R&D). Also, long immigration waits for Chinese citizens and growing anti-Asian discrimination, particularly during the pandemic years, has made life in the United States unwelcoming and unsafe for many Asians (Rathi and Lu 2023). However, 88% of international PhDs from India indicated intent to stay despite long immigration waits, perhaps due to the disparity in STEM research and career opportunities in India. Also, due to relative fluency in the English language in postcolonial India, many Indians have thrived professionally in the United States despite navigating a complex path to immigration. The data also indicate increased intent to stay in the United States among PhDs from Turkey and Iran, likely due to the unfavorable political and socioeconomic climate in their home countries.

Factors Influencing the Global Mobility of Scientists

In a *Nature* survey of 2,300 scientists worldwide, respondents were asked to rank the importance of different factors informing decisions to work abroad (*Nature* 2023). Increased quality of life and higher salary ranked as the strongest incentives for migration. Respondents ranked authoritarian political systems and fewer personal freedoms as the most significant barrier to migration. Difficulty obtaining work visas was ranked by 60.1% as a very significant or somewhat significant barrier to migration. These results align with the anecdotal comments and lived experience of many international PhDs in the United States who explore global careers.

KETAN'S STORY: FROM BIOMEDICAL PHD IN THE UNITED STATES TO SCIENCE CAREER EDUCATOR IN CANADA   Ketan Marballi moved from India to the United States in 2007 to obtain a PhD in cell and molecular biology. He was interested in a teaching-focused faculty career, but securing such a job requires significant teaching experience (afforded by teaching-focused postdocs). However, there are very few teaching-focused postdocs open to international PhDs. Therefore, he took research postdocs with the intent of developing teaching experience. There are also fewer federal teaching fellowships for temporary visa holders, so he secured a teaching fellowship offered by the American Society for Microbiology.

He saw a creative path to becoming an educator in the career and professional development (CPD) field. But because of visa restrictions, he was unable to secure these positions. As a result, he was trapped in a loop of doing multiple research postdocs and worried about financial stability as postdoc positions typically did not include retirement benefits.

In the meantime, Ketan was making progress toward permanent residency through an employer-sponsored green card. But like other Indians, he was stuck in a long backlog to secure the green card after his petition was approved. That's when he decided to explore careers beyond the United States. Canada was a good fit as it is culturally similar and geographically close to the United States, with a distinct advantage. The Canadian immigration system is point-based with incentives for high-skilled STEM workers with advanced degrees. Another benefit of a point-based immigration system is its straightforward and transparent criteria, compared to the ambiguous and subjective criteria for US permanent residency. He applied for Canadian permanent residency in 2019 through an express entry federal work study program. He remarked, "After experiencing mostly negative immigration-related experiences in the US, I was very surprised that I was able to get permanent residency in Canada within six months of applying." After that, he spent three months in Toronto to learn more about the lifestyle.

However, due to its relatively smaller economy, Canada's job market is highly competitive, particularly in a niche field like PhD-level career development. So he returned to the United States in September 2019 and took a research staff position at a university. This time, over the course of a few months, he established the clear goal to enter the field of PhD-level career development. He spent that year building skills through participation in the Postdoc Academy, the STEMPeers community, and Canada's Graduate Professional Development Network group (the Canadian professional society for graduate-level professional development). He talked to five or six professionals in Canadian companies in a week and volunteered with Canadian organizations, where he met his future manager at Toronto Metropolitan University. He was hired as a career education specialist in that university's Faculty of Science in 2021, within three months of relocating and after one year of preparing for the CPD field.

When I asked Ketan about challenges in Canada compared to those in the United States, he mentioned that salaries are lower and the cost of living is high in cities like Toronto. Also, changing jobs is hard because of the competition. Many immigrants move to Canada but leave after a few years due to fewer jobs and lack of career mobility. Nevertheless, Ketan concludes that he is happy with the trade-off, primarily as his mental health has improved because of his immigration stability. He also commented that he was less anxious about traveling. Traveling on a US visa can involve uncertainty and discrimination, particularly for brown men. Stories like Ketan's leave me with mixed emotions. On the one hand, I am thrilled that Ketan found his way to his desired career path in Canada; on the other, I think of many international scientists in the United States who are still stuck in the post-doc loop, unable to secure their desired jobs due to visa restrictions and living with anxiety and helplessness. If you are one of them, I hope you find courage and hope in Ketan's story.

IRINA'S STORY: FROM NEUROSCIENCE PHD TO EDUCATOR AND ENTREPRENEUR IN JAPAN   Irina Filonova identifies as a global citizen who made several fresh starts in her career, in Russia, in the United States, and in Japan. Before finishing her undergraduate degree in Russia, she came to the United States on a three-month work and travel program and never returned. She started college again, studying philosophy and psychology. Fascinated by the brain's complexities, she started PhD training in neuroscience.

When she began her PhD training, she was the mother of a two-year-old son. Juggling parental responsibilities, caregiving, and graduate school was challenging enough; constant imposter syndrome didn't help. Moreover, she started graduate school during the 2008 recession, resulting in a financial strain on her family. They even lost their house. That left no room for intentional professional development. Consequently, although she wasn't keen on a faculty career, she did a postdoc. Her family moved to Texas for her postdoc at the University of Texas

Southwestern (UTSW). There she attended a career development seminar, where she met Natalie Lundsteen, director of career and professional development, who became her career mentor and helped her develop confidence. Irina started exploring career options. She realized that she loved the academic environment, as well as teaching, and gravitated to leadership activities. She became president of the Postdoc Association (PDA). As a postdoc leader, she attended the NPA Annual Meeting, where she met Sina Safayi. Sina catalyzed another fresh start for Irina. He had just returned from his stint at the Okinawa Institute of Technology (OIST), Japan, and mentioned an open postdoc specialist position to build professional development programs for scientists. The hustle of a PhD and a postdoc and financial instability in the United States had taken a heavy toll on her and her husband, so in 2017 they decided to move to Japan with their twelve-year-old son.

The immigration path in Japan was easy for Irina's family. On arrival in Japan, they obtained a residential visa (equivalent to the US green card) for three to five years. Residential visas also follow a point-based system, which incentivizes scholarship and publications. But salaries in Japan are lower than those in the United States, and there is a lack of diverse career options. She didn't think much about it as she enjoyed working at OIST, until it wasn't enough. When exploring a job change, she realized there were few viable career options. The language barrier was a significant roadblock, and the style of communication in Japan is indirect and very polite, making interpretation harder. Unlike in the United States, in Japan people tend to stay in the same field and have linear careers. For a dynamic professional like Irina, this was a hindrance. By now she was burned out; the decades of hustle had caught up with her, and the pandemic made it worse. She took a sabbatical and participated in a meditation retreat in Thailand.

Following the sabbatical, she taught as an adjunct professor of biology at the University of Maryland's global campus. She realized she loved individual coaching and completed coaching certification and leadership training. Eventually she started her own coaching company and became an entrepreneur, seven years after moving to Japan. Irina doesn't know if she wants to stay in Japan long term. The family is open to relocating after their son finishes high school.

SHYAMTANU'S STORY: FROM GERMANY TO THE UNITED STATES AND THEN TO SWITZERLAND   Indian-born Shyamtanu Datta earned his PhD in molecular and functional genomics in Germany. After his PhD, he wanted to conduct applied research in the field of gene therapy to prepare for an industry career. He took a postdoc in gene therapy in Texas. During his postdoc, he learned about different careers for scientists and built skills through an internship in tech transfer, led the consulting club, and participated in consulting CASE competitions. He enjoyed business of science fields and could see a future in them. But he lacked a professional network as there are not many biotech and pharma companies in Texas.

Around this time, he came across STEMPeers and was amazed at the community-driven mentorship and professional development. He took a leap of faith and flew to Boston to attend the 2017 STEMPeers career conference. He remarked, "There was so much collaborative support at this altruistic community. Here I was trying to figure it out alone when I just had to ask for help, and there was so much support in STEMPeers." Through this network, he learned about the biotech ecosystem in Boston, but getting scientist roles in industry was challenging as he was on a visa for research scholars and lecturers in academia. He started looking for industry-like positions in universities in Boston. He got to know from the STEMPeers network that such jobs exist, although they are few in number. After a few trials and errors, he secured an academic-industry hybrid postdoc fellowship at Harvard Medical School that was partly funded by a multinational biopharma company. In addition, he got to work on other projects in collaboration with a small biotech firm. In two years, he had published some of the work and was ready to explore his next step. His visa posed challenges for industry roles. He was also aware of the long wait for Indians to get permanent residency in the United States and did not want to endure a prolonged cycle of helplessness.

He contemplated exploring jobs beyond the United States. His previous work experience in Europe made him think about moving to Switzerland, a country with a thriving biomedical industry consisting of biotech as well as big pharma. Parts of Switzerland are like Germany. And as he could speak some German because of the time he spent there, social integration would be easy in Switzerland. He applied for open positions on various companies' job portals in the EU. His combination of skills in a field of applied research that has a demand in industry made him uniquely qualified; he got interview calls from these cold applications, which led to a job offer.

Regarding life in the EU versus the United States, he echoed Ketan's and Irina's opinion that there was more flexibility and freedom in the United States due to more options. One needs to make some compromises in the EU, but that's a small cost considering the personal freedom it affords, particularly with regard to traveling regularly to see his parents. Shyamtanu is now working at another biopharma company in Switzerland and is happy with his life there.

THINGS TO CONSIDER FOR GLOBAL MOBILITY

If you are contemplating global mobility as an international scientist in the United States, consider the following factors.

Begin with Clarity of Purpose

Why do you want to leave the United States? Is your decision driven primarily by financial concerns, career advancement, a better lifestyle, family support, childcare, or mental well-being? Understanding the primary drivers will help you decide where to move. Start with self-reflection, then discuss your reasons with loved ones to reach collective decisions. Irina emphasized the importance of self-awareness when contemplating global mobility. She commented, "It's prudent to know your motivations." It serves us better to investigate motivations and our core purpose to approach decisions intentionally. A primary consideration will naturally be whether to return to your home country. Perhaps you would like to live closer to your family. It could also be driven by your career choice and experience living in the United States. A 2015 study surveyed international graduate students in physical sciences and engineering fields regarding their intent to stay or leave the United States after graduation (Han et al. 2015). Students who intended to pursue nonacademic or non-research-related careers indicated a 90% probability of continued stay. In contrast, students who aspired to academic research careers and believed that they would be treated better by colleagues back home had an 86% probability of leaving on graduation.

If you would like to explore moving to a new country, you need to take a comparative approach to determine which country will best suit your values and goals. You can start with a macroeconomic lens before focusing on local environments and individual preferences. In terms of macroeconomics, start with the viability of research careers indicated by the likelihood of available research jobs. One measure of a country's innovation potential is its R&D intensity or R&D investments as a share of overall gross domestic product (GDP). In simple terms, R&D intensity indicates the relative share of resources devoted to R&D in an economy, which likely translates to the proportion of STEM research jobs. AAAS publishes an annual report on global R&D intensity and the overall scientific research landscape globally. In the 2023 report, AAAS stated that China, Korea, Taiwan, and Israel showed the most growth in R&D investments in the past twenty years, followed by the United States (Zimmermann 2023). In terms of R&D spending, the United States surpassed all countries. When looking into R&D intensity, South Korea, Taiwan, and Israel surpass the United States although they are smaller economies. A large economy, China, aims to allot 3% of its GDP to R&D and already produces the most scientific publications and the largest scientific workforce. Therefore, when assessing the scope for a dynamic scientific career outside the United States, focus on countries with the most innovation potential, followed by countries that show the most R&D growth in your field of expertise.

Next, consider socioeconomic factors that will affect your quality of life. Talk to scientists in those countries to understand whether salary and benefits support cost of living expenses. You can also use COL calculators to compare living costs in global cities. Is there adequate social and medical support for families and children? If you are a single woman, you may also investigate safety in places of interest, as well as the societal attitude toward single women. For ethnic minorities, evaluate the acceptance level for multiculturalism. Avoid countries that have an increasing anti-immigrant climate, resulting in restrictive policies. Otherwise, you will encounter the same issues you faced in the United States. Platforms such as *Going Global* provide an overview of these socioeconomic considerations. Many international scientists also explore countries that are culturally like their home country or the United States for a smoother transition. Ketan moved to Canada as it was culturally similar to the United States, while Shyamtanu was comfortable moving to Switzerland because of its cultural similarity to and shared language with Germany.

Scientists contemplating global mobility often convey fatigue and fear of career transition, particularly because of the lack of knowledge about employability. For a reasonably smooth transition, you may consider jobs in a multinational company or at a global satellite campus of US universities. Shyamtanu created a bridge to Switzerland through a job in a multinational biotech company. Irina, on the other hand, took advantage of OIST and the University of Maryland's satellite campus in Japan. These organizations and institutions will provide a familiar culture cushion in a new country. There are a growing number of global talent entry programs for STEM PhDs in Australia, China, Canada, the EU, South Korea, and the United Kingdom, among others, that include diverse research fellowship opportunities.

Platforms such as *Euraxess* developed by the European Commission, *DAAD: German Academic International Network*, Talent Mobility Fund by

TABLE 2

Prominent Global Talent Programs for International
STEM PhDs

Program	Country
Global Talent Visa Program	Australia
Federal Skilled Worker Program	Canada
1000 Talents Program	China
German Academic International Network (GAIN)	Germany
Horizon 2020	EU
Brain Return 500	South Korea
High Potential Individual Visa Program and UK Horizons Program	United Kingdon

SOURCE: Adapted from Han et al. 2015.

Renaissance Philanthropy, and *Going Global* provide comprehensive resources for global mobility, particularly fellowships, career explorations, and job searches, as well as professional norms and cultures in different countries.

Also consider your goals when comparing a career in the United States to other countries. If, like me, you tend to change fields, leading to a nonlinear career path, you would want to move to countries that support your career adventures. Moving to countries that honor tradition and linear advancements in career trajectories and have a conservative employment and workforce culture may not be a good fit for your personality, as Irina experienced in Japan. That said, Irina found independence in entrepreneurship, when she began her coaching consultancy. In line with her overall trajectory, Irina offered an important piece of advice: embrace serendipity. She believes that people can get trapped in goals and strategies. While planning is good, one should not get too caught up in it. She said, "You can plan all you want, but the future doesn't need to comply. For instance, my career is a series of random events. Be flexible, and take a leap of faith. Life in a new country will suck at first. But eventually you'll find rhythm and know how you truly feel. It's worth investing one or two years of your life on this adventure. In the wider arc of your life, it's nothing." Even after a few years, if the new country doesn't suit you, you can explore returning to the United States. If your employer has a base in the United States, there are visas for intracompany transfers (L-1 visa) for managers and executives.

Finally, if salary growth is an important consideration, all data and anecdotes point to the fact that salaries and growth opportunities in the United States provide more benefits in the long term once you can secure your immigration pathway.

Approaches for Global Mobility

As in career exploration in the United States, you should talk to other international PhDs with a shared background and field of work who migrated to another country. This is especially crucial. Those with firsthand experience can offer a clear perspective on social and professional life in a different country—its culture, quality of life, language norms, and degree of acceptance of immigrants. I hope Irina's, Ketan's, and Shyamtanu's stories help you realize that there are many international scientists whose experiences can help you. Shyamtanu emphasized the power of networks to provide exposure to career options and broaden your imagination of possibilities. Irina is also a big believer in finding your global community. We both had a moment of deep validation while discussing the point that internationals have a way of finding our global community. As global nomads, we are curious about cultures. When internationals meet, we ask about each other's homes and cultures.

Culture is a curious thing, though. You can learn about different cultures from people to an extent. Unless you live in a place, you cannot truly discover the culture. One approach is to move to a new country for a short period before making long-term decisions. That is what Ketan did when he lived in Toronto for three months before relocating permanently. A short-term job or postdoc will enable you to experience the culture, support systems, and daily life of residents. Irina, Shyamtanu, and Ketan all gave this advice. Shyamtanu also commented, "It's easier to move within multinational companies (MNCs) with robust infrastructure, options, and shared global workplace cultures." MNCs make dynamic global mobility between their offices easier; many of his American and international colleagues had returned home after a few years in Switzerland. As multinational companies typically have a professional monoculture, you will have a familiar work culture while you navigate life in a new country.

Social integration is certainly easier in the United States, especially in cosmopolitan cities. To clarify, integration and assimilation are two different things. Integration encourages coexistence of diverse cultures within societies, whereas assimilation assumes superiority of the dominant culture, typically of citizens over immigrants. Social integration is harder in closed societies where assimilation is expected. Moreover, language barriers in non-English-speaking countries can be hard to overcome for immigrants, particularly without working knowledge of the local language. As Irina mentioned, even with a working knowledge of Japanese, cultural contexts and nuances in communication may be lost in translation. Therefore, it's prudent to be informed about the desired level of fluency in native languages to navigate life in different countries.

Every society has its own sets of expectations, professional standards, and career norms. You need to adapt to different styles of CVs and cover letters. When I asked Shyamtanu about differences in the job search culture in Europe, compared to the United States, he commented that he found a happy medium as MNCs have a shared culture across their global offices. That said, he avoided too many quantitative and individualistic attributes in his application to make it more suitable to European conventions. Also, professional communication styles in some countries may have a preference for directness, whereas other societies may find that rude. Professional fields also have their own norms that can transcend borders. Thankfully, the professional cultures in the sciences have some universally shared qualities. Career advancement, self-promotion, and self-advocacy, which are common-place in the United States, may be frowned on in collectivist societies.

I hope to have demonstrated that you do not need to limit your ambitions and sense of adventure. The world remains your oyster; if you can move once, you can move again. Start with clarity of purpose, and gain knowledge of global options, socioeconomic and political conditions, and quality of life to determine your desired destination. Rely on people who can provide firsthand knowledge of life there. Look into the job search and application process in other countries via resources and connecting with other scientists who made similar moves. Take a leap of faith to experience life in a different country for a few years. Ultimately, global careers broaden options for international scientists, especially those from India and China, who face steep immigration challenges in the United States. Unhindered by restrictions, international scientists can exercise choice and get a true measure of their competitiveness in the global marketplace.

KEY HIGHLIGHTS

- International scientists are more likely to migrate early in their careers.
- Increased quality of life and a higher salary are the biggest incentives for migration, while political systems and lack of personal freedom are barriers.
- When exploring global careers, investigate STEM R&D intensity (R&D investments as a share of overall GDP) to gauge the career landscape and innovation potential of a country.
- Talk to international scientists who have migrated or browse resources to compare socioeconomic factors such as salary, cost of living, job competitiveness, cultural integration, and language barriers.
- You can test the waters in a new country by moving for a postdoc or a job at a multinational company before deciding to migrate long term.

Beyond a Model Minority

The Humanity of International Scientists

CHAPTER TEN

The Mental Well-Being of
International Scientists

International scientists, like many high-skilled immigrants, are viewed as a model minority who get the job done by overcoming all odds. We are hailed for our productivity, for STEM innovation and economic growth, for creating jobs and filling critical labor gaps. Such rhetoric is certainly helpful in maintaining goodwill regarding immigrant advocacy. However, these broad generalizations do not consider the dehumanization and mental health burden of immigrants. We are cited in economic and workforce metrics without properly acknowledging that we are complex human beings. Immigrants amplify this discourse for self-advocacy purposes. We are praised for our work ethic and resilience, and we certainly take pride in it. However, this resilience is not a choice but a survival mechanism in restrictive systems and exclusionary environments. Immigrants are forced to be resilient due to frustrating and convoluted immigration systems, inequitable access to opportunities, trauma from discrimination, workplace microaggressions, bullying, and sometimes outright xenophobia. In addition to celebrating international scientists' resilience, we can also invest efforts to make an inclusive environment for immigrants where resilience isn't a survival trait.

In this chapter, I bring attention to the mental toll on international scientists who are separated from loved ones for decades because of travel restrictions, who endure systemic dehumanization and workplace bullying to maintain their immigration status, and who regularly feel inadequate due to language barriers and feel invisible despite their talents and acumen. I ask colleagues and managers of international scientists to respect, support, and advocate for their international colleagues in meaningful ways and call for inclusive and humane systems. Finally, I encourage international scientists to give themselves grace and seek help and support for mental and emotional well-being.

While the scientific enterprise relies on immigrant labor and celebrates its global culture, it falls short in meaningfully advocating for international scientists. Worse, international scientists are expected to take care of their existential challenges outside of work hours. Navigating the immigration maze and efforts to maintain visa status are strenuous and stressful. Instead of supporting and empathizing with their trainees and employees, many managers and faculty advisers see these concerns as a hindrance. I have heard countless international graduate students and postdocs mention that they must take care of their immigration paperwork outside work hours, preferably on weekends. In some cases, they feel hesitant to bring it up to avoid inconveniencing their colleagues and supervisors. I speak from experience. Constantly decentering the self adds up in the long run and breeds fatigue and resentment in equal measures. I sincerely encourage you to strive for holistic lives that give equal weight to joy and well-being as much as ambition and career goals.

THE UNDERLYING SACRIFICES OF IMMIGRANT LIFE

On top of the challenges in scientific life, international scientists sacrifice long stretches of time with their families and miss milestones and family functions as going home regularly is not feasible. Harmit Malik, professor and associate director in the Fred Hutch Basic Science Division, mentioned that homesickness can be especially hard for internationals in the beginning. He advised international scientists to create a sense of community early, whether with peers from their country or with those with shared interests. While this advice may seem contradictory as spending time with peers from one's own country may present fewer opportunities for cultural assimilation, it is important to strike a balance between familiarity, shared experience, and new cultural experiences in the early days.

For many of us, saving money and time to visit family once a year is an ordeal. Planning it around a family wedding or a milestone while juggling research deadlines is another difficult project. As a result, we miss a lot of important family moments and become distant, even outsiders. We miss the context in conversations. I missed the birth of my niece and most of her milestones since, weddings of every cousin I grew up with, and, worst of all, my parents' deaths, last rites, and funerals. When I finally was able to go home, a few months after my parents' passing, the grief of their absence at home hit me hard. The rest of my family in India were in a different phase of grief, and I felt guilty yet again for expecting them to honor my out-of-sync emotions.

Many international scientists mentally prepare for days before requesting leave to visit home. I know of many instances when they were made to feel guilty about this basic human right of visiting family. There's always a manuscript deadline or work thing that we are asked to accomplish remotely while visiting home. You feel (or are made to feel) worse if you get stuck for long periods due to visa-related backlogs. Some research advisers and managers can be downright selfish and unfeeling instead of supportive. They can't see beyond their deadlines. They focus on the hindrance to their lives, with much less consideration given to the international students or postdocs going through the uncertainty of visa issues. I have heard many such stories, more than I would like. Whether it's absence for visa reasons or medical reasons, many international scientists work through the uncertainty owing to their bosses' expectations, even when they are not required to do so legally.

We international scientists remain focused on our academic and career goals at the expense of our mental health. We internalize the belief that we can power through any adversity, but over time the pressure builds and the outcomes don't measure up to the collective cost. During our PhD training, we keep striving toward the dissertation defense, and when the defense is over, it's disappointing how unremarkable it was. It is at this time that many PhDs experience a mental health slump. We wonder whether it was worth the sacrifices. Do our structures and environment need to be so exhausting and inhumane that no accomplishment feels worth it because of what it takes to get there? Would it be so bad for systems to be humane so that international scientists can live whole lives?

THE UNYIELDING HUSTLE FOR EXISTENCE

Before you know it, you are busy with the next thing. You hardly have time to celebrate or process disappointment or properly reflect on your journey to a milestone or even take time off (because of the visa clock) before you are hustling again. I wrapped my PhD in December in Athens, Georgia, and started a postdoc in New York City in January. I was planning my move in parallel with preparing my dissertation for final submission. On top of that, my whole family (parents, brother, sister-in-law, and niece) came from India to celebrate my commencement. So, while juggling dissertation revisions and my move to New York, I also planned their visit to include some memorable touristy things in the United States as they were spending a lot of money to travel across the world. It was all wonderful, and they loved every bit of that trip. But I was simply going through the motions, checking off my to-do lists. I relocated to New York in the middle of snowy January and started my postdoc right away. As I mentioned earlier, I haven't taken any substantial breaks between PhD, postdoc, and jobs over the past seventeen years in the United States. I am exhausted at the thought of additional planning and dealing with more immigration bureaucracy for a month break between jobs.

Most international scientists feel this exhaustion after big moves and transitions. Jenny Schneider earned her PhD in developmental biology and embryology in Israel and moved to the United States for postdoc training in neuroscience in New York. Once she moved, it didn't feel like the big moment she imagined because of the rigors of setting up life in a new country. She recalled, "After putting so much effort on moving to New York, you would think that when the day finally came, it would be a gratifying moment. But it wasn't like that. There's so much to do when you move to a new country. You're busy opening bank accounts and trying to figure out how to get your apartment keys without a social security number as you can't make check payments. But after the initial jitters, I found home and comfort in New York. It felt like I made the correct decision." Transitions are also hard due to cultural adjustments. While Jenny felt a sense of belonging in New York, she experienced the culture shock of academic scientific research in the United States. She mentioned that while scientists in Israel work hard, there is a culture of care. She recalls her research experience in New York as a "do it or get out of the way" hustle. It was dehumanizing; there was a lack of grace, respect, and gratitude for one's labor. Moreover, the culture of overwork results in erosion of a community spirit as no one has time to help someone else even if they wanted to. In the United States she felt like a machine producing data, not a scientist. In Jenny's perspective, scientists in training in Israel are no less productive than in the United States but endure less disrespect and have better mental health because of the work-life balance. Indeed, graduate students and postdocs are expected to work evenings and weekends. The work culture is worse in labs, which are populated primarily by international scientists. I am aware of labs that host group meetings on weekends with the expectation of weekend work. As Gary McDowell, CEO of Lightoller LLC, told me, "I have found that domestic scientists don't put up with the kind of conditions that internationals endure."

WORKPLACE ABUSE

Due to power imbalances and structural barriers, we are prime targets of gaslighting, discriminatory behavior, and bullying. As many of us don't know any better, we may not have a frame of reference about standards, so we may not even know when more is being asked of us than our domestic peers. Many of us ask our domestic colleagues; we compare whether we are treated differently. We may even be aware of incidents of bullying and discrimination. But can we say anything without fear of consequences in a power imbalanced system where your ability to continue working (or studying) in a country is on the line? Abusive bosses and research advisers use one's immigration status as a bargaining chip implicitly and, sometimes, directly. And persistent discrimination, toxic behavior, and gaslighting can harm the self-esteem of international scientists. We stop recognizing our value. We are made to feel inadequate and less talented. Worse, we may internalize it, compounding our imposter fears. We quietly endure toxic behavior while we are expected to be thankful for access to reputable institutions given our perceived seemingly limited talents. Consequently, misuse of power goes unquestioned and vicious cycles of abuse persist until a bystander or ally speaks up.

How to Combat Gaslighting and Navigate Risks

Many international scientists endure toxic workplaces as they feel helpless due to immigration rules. First, we may not fully understand visa-related limitations and whether a risk is real or inflated. Second, we aren't aware of our legal options. To effectively withstand gaslighting, I suggest that you gain knowledge of employment laws and immigration rules. If you must spend money on something, let it be on credible immigration counsel. Try consulting with an independent immigration counsel, in addition to the one offered by your employer. It will pay off in the long run as credible immigration information will inform your decisions and you will not be operating out of fear. Third, become familiar with employment laws in your city and your state, as they may be different. Seek guidance on employment rules from staff in graduate, postdoc, and international offices, particularly about student and worker status, postdoc contracts, and notice periods. Many international scientists are scared that they may not get sufficient notice to change jobs. Through online communities of international scientists and the trainees I advise, I am aware of wretched stories of principal investigators giving very short notice to international postdocs. That is why it is important to know about terms of employment early on. I learned from a university staff member about making a strategic decision about the last day of employment; for example, if you leave during the first week of the month, your health insurance for the rest of the month will likely be covered by your past employer. That way, you will have a cushion for a few weeks. And as we usually do, rely on the international community for support. But do it proactively, not after you are in crisis.

Risk-taking is inherent to international scientists' lives because of immigration precarity. It comes down to one's risk tolerance. Risk tolerance varies based on circumstances. Chinese and Indian citizens are most vulnerable as their path to permanent residency is long. The visa precarity makes changing jobs harder due to the sponsorship burden. Moreover, international scientists are particularly vulnerable during an unstable employment climate. Adding to that, international scientists with families are responsible for dependents as well as their partners' careers, which lowers their risk tolerance. They are more likely to persist in suboptimal workplaces due to job stability and immigration prospects. Despite my visa precarity, I could risk changing my career path as I am single and without dependents. It is also a matter of personality; some people are more comfortable with risks than others. But you must honestly evaluate the trade-off regarding your mental health and how it may be affecting your relationships and performance.

These are hard decisions because sunk costs can have more weight for internationals. You may not want to give up on unfinished papers and projects that involved years of effort and are important to your immigration portfolio. That said, please think in the long term. If there is a realistic path to permanent residency within a defined period, you could stick it out for a few more years, but if you don't have a concrete agreement with your supervisor about long-term visa sponsorship and a path to residency or project timelines, you need to step up and have hard conversations. At the very least, you will know where you stand and can assess your options. Will you take the loss and move to a different workplace where you are respected and will, likely, thrive? Be wary of bosses who do not commit to anything concrete and make circular arguments. That's a definite red flag; you won't thrive in places that lack honesty, mutual respect, and trust.

Jenny realized she had an unhealthy work life and decided to explore other career options. She sought help from career advisers, participated in skills workshops, and volunteered outside work hours. Two factors influenced her career direction. Frustrated with the postdoc experience and the scientific research enterprise and inspired by the help she received from career advisers who were former bench scientists, Jenny found her path into PhD-level career and professional development through an administrative postdoc position at Thomas Jefferson University (TJU). Now Jenny leverages her data analysis skills to assess the postdoc experience and career outcomes in order to make improvements in scientific training, provides data to support institutional NIH grants, teaches a course on professional development, and guides scientists in figuring out their paths at TJU and beyond. Jenny's mental health has vastly improved as she is in a healthy work environment. Ever a scientist, Jenny continues to harness data-informed ways to make systemic change in scientific research. If you take any lessons from her experience, please listen to your conscience and remember why you became a scientist in the first place. If you find no joy in doing research in your current environment, why prolong your misery? Who does it help? Certainly not you and maybe not even your colleagues and boss. So do some risk analysis, seek help, cut your losses, and work for a healthier future.

GLOBAL CRISES AND THE STRUGGLE
WITH IDENTITY

The ongoing cycle of unprecedented times, with geopolitical and climate crises, a pandemic, and international conflicts, has added mental strain to the lives of international scientists. Moreover, our lives in the United States can be dissonant with our native identities and the struggles of loved ones back home. Our part of the world may be upside down, and we may be in internal turmoil. But our workplace is calm. So we compartmentalize and show up projecting composure to sync with our work world. We can't bring the upheavals in our inner lives and our home country to work. I recall that when India was in the thick of the COVID-19 crisis, things were relatively stable in the United States. I felt so helpless watching the mass death back home. Thankfully, I found a community of Indian scientists and allies in the United States and Canada who formed the group India COVID SOS to provide aid and assistance through knowledge and resources in science and medicine.

The international scientific community is a multicultural haven. But international conflicts make our lives very hard. The current war between Russia and Ukraine as well as the Israel-Palestine crisis have been very challenging for the international academic community. The visa regulations make it even more complicated. Internationals have limitations on the right to voice political opinions or to protest due to visa stipulations. We self-censor; we try to hide in the crowd so that we don't stand out as supposed threats to anyone by simply existing. To make matters worse, we are forced to respond to questions about political beliefs in green card applications, visa interviews, and so on.

Due to geopolitical tensions in the United States with countries such as China, Russia, and Iran, scientists from those countries also navigate perceptions of being national security risks and cannot visit their families back home for decades. I remember a heartbreaking moment when I attended a memorial for an Iranian graduate student. The Iranian student community hadn't traveled home for many years, so they had formed a familial bond with each other. The remembrances at this student's memorial reflected a profound tragedy beyond the life we were memorializing. Many Iranian students, postdocs, and researchers hadn't visited their families since they had stepped foot in the United States, spanning a decade for some. Whatever one's sociopolitical alignment, on a human level, it is so unfair. I wish I could better explain the feeling of sheer loss and dehumanization that many international scientists endure for no fault of their own.

Everything adds up, whether it's compartmentalizing, self-censorship, shrinking oneself, or living partial lives while enduring long-term distance and disconnection from loved ones. The compounded effects of these existential struggles and forced resilience contribute to declining mental health. On top of that, many of us come from cultures where mental health isn't openly discussed or is stigmatized. We don't think about actively seeking psychological help and have a complicated perception of therapy. While there have been some improvements, many international scientists still believe they can soldier through adversities on sheer will and determination without seeking help or therapy. Another factor is that many of us grew up in cultures that center communal healing. We do not sit in a room with a therapist to process our feelings; we share our burdens, our grief with our loved ones and communities. But wherever you fall on the spectrum of comfort with seeking help, you can still take small steps to support your well-being. Make time in your routine for restorative activities that bring you joy and peace. For me, it's going on walks with my favorite playlists, sleeping in on the weekends, or even a brief morning yoga routine. Reconnect with your hobby, find time to catch up with friends and community, or make time to do nothing at all. Ultimately, finding joy in other activities can remind us that we are more than our work and immigration status. It is crucial to move beyond our professional and immigrant identities, particularly through challenging times. Give yourself space to heal, to rewire for the next sprint. If you are an introvert, you can journal. Journaling, or the exercise of processing by writing, is powerful. Also, communal healing can be a good substitute if you are deeply uncomfortable with therapy. You can join support groups; start with listening if you are hesitant to share. Look for affinity groups or communities of support with shared identities. You will find others who share your experiences and fears. You won't suffer alone. That said, please find the courage to seek professional help if you are struggling. Many universities provide access to counseling and psychological services for graduate students and staff as part of their health coverage.

My final note is for peers, friends, and colleagues of international scientists. We all rely on our communities for support. In the absence of their families, your international colleagues may need encouragement, allyship, and support from you. If you are an American by birth or naturalization, I hope you actively support your international colleagues. Don't assume they are fine because they seem calm on the outside or don't ask for help. Take them out for coffee and get to know them; ask them how they are feeling. If you are a bystander to unfair treatment of international colleagues in the workplace, please speak up, advocate, or direct them to help and counsel. At the very least, you can help uncover the hidden curriculum for your international peers. Make them aware of differential behavior or expectations. If you are a supervisor or manager of international scientists, I have an upcoming chapter for you.

CHAPTER ELEVEN

International Scientists Are Not a Monolith

We talk about "international scientists" as a monolith, a homogeneous population. However, our experiences and challenges can differ greatly due to diverse backgrounds, identities, and circumstances (socioeconomic status, country of origin, race/ethnicity, gender, non-native English speakers, field of study, educational pedigree, prestige of graduate and postdoc institutions). For instance, international scientists from developing countries face additional disadvantages due to layers of systemic barriers. In this chapter, I encourage you to understand and address the varied experiences of international scientists and identify the compounding challenges faced by groups of international scientists that contribute to their marginalization. While tailoring guidance and support for international scientists is a first step, mentors and managers will need to personalize support based on the disadvantages and disparities their international mentees or employees face, resulting in marginalization.

MARGINALIZATION FACTORS

While advocating for support of international scientists is critical, treating that population and its needs in broad strokes is unproductive.

International scientists can have vastly different experiences with wide disparities due to several factors that can contribute to marginalization and hinder success. For the sake of simplicity, I refer to these as "marginalization factors." Marginalization factors include shared drivers that leave domestic minorities as well as international scientists bereft of agency and power, such as gender, race/ethnicity, first-generation status, and socioeconomic class. These factors can also contribute to disparities in the lived experience of international populations, in addition to historically marginalized US citizens. Further, specific marginalization factors, such as country of origin (or passport strength), immigration status, degree of fluency in English, and social and economic stigmas borne by those from Global South (lower-middle income [LMIC]) nations, have differential negative impacts on certain international populations. In the following sections, I describe how these marginalization factors create disparities in the lived experiences of international scientists. Fair warning: These topics may make you uncomfortable due to the geopolitical and historical dimensions that continue to persist in modern systems and society. But they need to be discussed as they affect the lives of internationals, sometimes in a dire manner. I won't speculate on the underlying reasons (some are obvious) for these disparities. For one, I am a natural scientist (not a social scientist) and lack the appropriate expertise or language to connect historical contexts with current biases and barriers. I can, however, bring attention to the tangible ways in which norms and systems create material disparities in the lives of international scientists through observation and insights gained from the stories of fellow internationals.

Country of Origin

The popular narrative has it that the American Dream is merit driven and favors the talented, curious, and brave. While somewhat true, one may argue that country of birth plays a significant role in the success of international scientists in the United States. The strength of our passport determines how bumpy (or smooth) our life will likely be. The global passport rank determines strength on global mobility scores (Passport Index 2024). The top thirty-five passports with high global mobility scores are issued by countries in Europe and North America, with a few exceptions. Given that the scientific enterprise is highly global, with many international conferences, ease of global travel is key to accessing knowledge and professional networks. International scientists with weaker passports have access issues due to travel restrictions and hurdles such as needing visas, money, and additional time to plan conference travels (Chugh and Joseph 2024). Chinese and Indian passports have low mobility scores, resulting in limited global travel for these citizens. This is important as China and India are the top two countries of origin of international scientists in the United States. On a personal level, travel restrictions hinder one's ability to meet with one's family regularly, as I discussed in chapter 10. Fraught geopolitical relations with the United States adds additional strain to international scientists from certain countries. For instance, international scientists from China or Russia or Iran (or broadly, the Middle East) have to regularly navigate social stigma, bias, and scrutiny as potential national security threats (Xie et al. 2023). These scientists also face more xenophobia.

Immigration Status

In addition to citizenship privilege, experiences of international scientists can vary depending on the type of visa they hold. I mentioned earlier that many J-1 visa holders have a two-year home residency requirement, and the waiver process is long. There are limits to the number of H-1Bs (work visas) offered annually. H-1Bs can be extended for a maximum of six years (three years at a time). Even international scientists working in the United States must go abroad to get their visas stamped at a US consulate or embassy. Traveling abroad on an expired status can be stressful. If the consulate decides to do a background check (administrative processing) during the visa interview, one may be stuck outside the United States waiting for their visa for over a month. If this sounds bad, imagine going through it every three years. I went through administrative processing every time I went to apply or renew my student (F-1) visa or H-1B work visa during my PhD and postdoc (three or four times over ten years) as I was doing scientific research. Therefore, I would always schedule visa interviews in India, no matter how long the waitlist for an appointment. At least I was stuck at home while the background check was ongoing.

As discussed earlier, pathways to permanent residency (a green card) can be long for many international scientists (over a decade for Indian scientists). While waiting their turn for a green card, most international scientists can renew their work visas. In theory, it doesn't sound bad, but imagine spending time and money on petitions, scheduling interviews, and likely getting stuck in administrative processing delays abroad every three years while waiting in a queue for ten or more years. The worst part is the unpredictability; one's position in the green card queue is arbitrary, and it can move forward or backward for years at a time. Curiously, it is faster to get a family-based green card than an employment-based one in some scenarios. For example, an Indian citizen is more likely to receive a green card faster on marrying a US citizen. It's almost a customary joke that many attorneys make during immigration seminars. After outlining the immigration maze, they end with, "Or you could marry a US citizen." I was on the receiving end of this joke many times. I responded to it with my deadpan face, but internally I would cringe at the dark humor of the system. So much for being merit based. I hope you can recognize that success is also related to one's birth lottery. If you are an international, you are in better luck if your country of birth presents a stronger global passport rank or one where your wait for a green card (and naturalization) is shorter. Otherwise, here's hoping you fall in love with and marry a US citizen! Crass? Yes. My point is that not every international scientist's struggles are the same.

Family and Dependents

International scientists with families and dependents have additional risks to consider. The cost of living and childcare expenses are exorbitant and untenable for all early-career scientists with children, including international scientists. Adding other risks such as visa precarity and the fear of losing benefits even for a short time makes it hard for international scientists with dependents to be adventurous about careers. Given the risks, internationals with dependents are more likely to persist in suboptimal, even toxic workplaces.

Race/Ethnicity and Nationality

When it comes to internationals, race can feel like an inaccurate descriptor; ethnicity and nationality are better identifiers. Specifically, the interplay of race, ethnicity, and nationality shapes the complex and differential experience of internationals. White internationals can blend in easier in US society than Black and brown internationals. By no means do I want to minimize the historical and ongoing systemic and violent oppression of minoritized US citizens, particularly Black Americans. But to some extent, the systems and social stigmas that marginalize Black and Latino/a citizens also impact Black and brown internationals due to physical similarities. As a result, instead of solidarity, many Black and brown internationals resort to self-preservation by distancing (and distinguishing) themselves from Americans of color through class, qualifications, and skills status. White adjacency is systemically favored, which creates divisions between internationals and citizens of color and triggers a competition for systemic favors. Often coded as "professionalism," whiteness is also culturally and professionally favored, and many internationals try to game it to succeed, even if it means masking aspects of their ethnicities. Colorism adds another layer among Black and brown people that influences who can assimilate and claim white-adjacent advantages.

The lack of nuance in defining the identities of internationals is systemic; ethnic lines are arbitrarily drawn by the US Census Bureau. One example is the curious category "Asian," which lumps 60% of the global population into one category. Asians are very diverse, of course. They experience diverse types of stereotypes, stigmas, and xenophobia. Due to historical and political contexts (i.e., World War II, the Cold War, the red scare), East Asians experience a specific flavor of xenophobia, not shared by South Asians. Post-2016 political scare tactics about China and the post-COVID-19 era has made East Asian marginalization worse. South Asians faced a different kind of bias and stereotype in the past (the head-bobbing, exaggerated Apu of Simpsons stereotype), but the confident, English-speaking STEM leaders and South Asian elites flaunting Indian food and culture have created a weird model minority niche within the South Asian population. Internationals from the Middle East and Arab-Muslim world have had it really bad since 9/11; I lack words to explain the level of dehumanization they face. In addition, the ongoing wars in the Middle East add further complexity to their marginalization. All that is to say, awareness of racial-ethnic and geopolitical contexts that affect international scientists is critical when supporting them. A Chinese scientist is probably masking different kinds of trauma than is a Syrian scientist.

Socioeconomic Conditions

THE GLOBAL SOCIOECONOMIC DIVIDE   There is a specific shared history among many Black and brown internationals: colonization. Colonization has created disparate economic conditions for generations of internationals from postcolonial nations. Moreover, we exist in systems historically designed by white men that favor rich nations, the very nations that either directly colonized or benefited from colonization. This disparity is recognized in the Global North-South divide.

THE GLOBAL NORTH VERSUS THE GLOBAL SOUTH   According to the United Nations Conference on Trade and Development (UNCTAD), the Global South broadly comprises nations in Africa, Latin America, the Caribbean, Asia (excluding Israel, Japan, and South Korea), and Oceania (excluding Australia and New Zealand) (UNCTAD 2023), typically LMIC countries with large populations that bear the economic brunt of prior colonization and oppression. One practical economics metric is the conversion rate of the US dollar (USD) compared to one's national currency. Internationals from LMIC nations incur an additional financial burden due to the steep conversion rate of US dollars. Compared to local wages, internationals from many LMIC countries make substantial investments to set up life in the United States. For example, the graduate school application fees seemed especially high to me when converting US dollars to Indian rupees (INR). I spent a good portion of my monthly salary to apply to up to five US universities. I had to strategize well to ensure the success of at least one of the applications. I moved in 2007; the USD/INR rate has doubled between 2007 and 2024. The stakes are even higher for internationals from LMIC nations due to the steep financial cost.

Socioeconomic and cultural systems harbor citizenship privilege, favoring internationals from the Global North (Chugh and Joseph 2024). Citizens from the Global North are more trusted in the United States and earn empathy and respect not easily available to citizens from the Global South. There is a specific implicit bias associated with the assumed cultural and intellectual superiority of the Global North. This also plays out in the microcosm of scientific labs. International scientists from the Global South are more vulnerable and likely face greater bullying and abuse than their European colleagues. Gary McDowell, a scientist from the United Kingdom who came to the United States for his postdoc and is now a CEO and academic consultant, conjectured that American scientists are less likely to endure bullying and bad working conditions compared to international scientists. During our discussion, Gary speculated on the underlying bias in academic discourse about the "lack of good quality postdocs and postdoc shortage." The trends tell a different story: the overall number of postdocs in the United States has not decreased significantly; there are fewer domestic postdocs (Langin 2024). There are also likely fewer foreign postdocs from European countries due to the antiscience bias and hard-right political turn in the United States since 2016. The recent growth in foreign postdocs is accounted for primarily by Asian countries. I wonder whether the "lack of quality" discourse points to this fact.

The experiences of internationals from the Global South, grouped together ethnically, also differ. Culturally and socioeconomically dominant nations in the group can mask the experiences of other nations. If we consider the South Asian category, Indians tend to dominate the narrative more than do Nepalese, Pakistanis, or Sri Lankans. This may be due in part to the fact that some Indians may have relatively greater socioeconomic privilege, which is overlooked as a result of a lack of nuance. The experience of Nepalese may differ greatly from Indians, but those contexts find very little visibility.

INDIVIDUAL SOCIOECONOMIC AND CLASS PRIVILEGE   The socioeconomic status and class privilege of individuals within the same ethnic or national groups as well as across groups also differentiate experiences. Internationals from upper-middle-class or high-income families have a financial cushion that assists in the transition and aids in mental well-being. Setting up life in a new country is expensive. From graduate school applications, GRE fees, visa fees, and airfares to deposits on apartments and daily expenses, there is significant upfront cost before one sees the first paycheck or stipend. Many international scientists and their families invest much of their savings or take out loans to cover these setup expenses. For many in the Global South with English as a second language (e.g., India), education in a primarily English-language school can be more expensive. If, like me, you speak English fluently (and without a regional accent) compared to your peers, it's likely a sign of socioeconomic class privilege and will set you up for success in the United States. Countries like India have other problematic social hierarchies, such as the caste system. Upper-caste Indians have greater access to social and economic privilege at home, which drives access and likely success internationally too. We may lack relative privilege in the global context, but within our own peer groups, some of us may have other privileges.

English-Language Fluency

As English is the accepted language of science, international scientists who can communicate fluently in English have a better experience in the United States. Non-native English speakers face additional challenges and marginalization (Carlsson 2024). They can struggle with navigating scientific presentations, manuscripts, and networking. In addition, they can miss cultural references and colloquialisms often used during socializing. International scientists from postcolonial countries typically learn English as their first or second language, providing an advantage in the United States. Citizens of countries that have easy access to US popular culture and entertainment will experience less culture shock and find the transition easier.

One of US society's persistent flaw is equating competence, skills, and intelligence with English fluency and communication skills. A good communicator can get away with a lot in any domain. Even the US presidential elections rely on debating and communication skills. Confident orators are automatically called good leaders. Our society needs to be more inclusive and expansive in appreciating the resilience and determination of non-native English speakers (Carlsson 2024). They work harder to interpret and communicate their thoughts and work in a different language. Having an accent is another dimension that generates bias. Internationals with strong ethnic accents are judged harshly. That said, there is differential treatment in this regard, which is yet again associated with white adjacency and the Global North-

South divide. For example, compare the reception of a scientific talk given by a French scientist and a Chinese scientist. Both may speak English with strong regional accents, but one may find more acceptance and patience than the other. The cultural reverence in the United States for the French language is not extended to Mandarin or Spanish or Hindi or Arabic. This introduces nuanced differences in treatment given to non-native English speakers based on which part of the world they come from. Some accents are considered cute or sophisticated; some are not. Worse, some languages (or accents) may be equated with danger, triggering more xenophobia than others. Let's look at these social biases and how they impact the experiences and confidence of equally talented international scientists.

Prestige of Graduate and Postdoc Institutions

Whether or not academics admit it, academic pedigree plays a role in assessing credentials. Letters of recommendation from eminent scientists as well as the reputation of institutions and faculty afford advantages to candidates in securing scientific positions, particularly in academia, whether it is graduate school, a postdoc, or a faculty position. In general, most of these eminent scientists and prestigious institutions are in the United States, Canada, or Europe. Therefore, international scientists who gain early access to these institutions typically have an advantage over internationals who move to these institutions for a postdoc or later in their careers. In addition, early access to these institutions also brings resource advantage. For instance, international scientists who came to the United States for graduate school likely have more papers in high-impact journals than do foreign-trained PhDs who move to the United States for postdocs. Typically, scientists in the Global South struggle to match the resources and funding for cutting-edge, high-throughput, and glamorous research and expenses required to publish in journals like *Science, Nature,* and *Cell.* Therefore, earlier access to resource-rich academic elite institutions with prominent faculty also creates disparities within the international scientists' experiences and likelihood of success.

Hidden Curriculum

Like first-generation college students in the United States, international students navigate a hidden curriculum, the cryptic rules of success. They must quickly learn about the hidden academic rules to catch up with their domestic peers. The hidden curriculum spans many professional areas: academic work cultures, conferences, oral and written communication styles, liberal arts (even intricacies of the "scientific method"), benchmarks for success, scientific publishing, networking, job searches, and interview processes, among others. There is a hidden curriculum associated with every topic discussed in this book. The hidden curriculum does not affect people equally, as it may create differential experiences among international scientists. Internationals who came to the United States for undergraduate or graduate school have a head start uncovering the hidden curriculum, compared to internationals who migrated later. Also, not all international scientists are first generation. A few may have scientist parents who have some understanding of the academic experience in the United States. That said, as internationals navigate the hidden curriculum in many areas, the advantage of prior knowledge in one area may not carry over.

Sanjana Mukherjee's experience is a good case study of the multifaceted nature of the hidden curriculum. As I mentioned in chapter 3, Sanjana is a professional fellow in global health policy at the O'Neill Institute for National and Global Health Law in Washington, DC. Sanjana came to the United States from India for doctoral training in microbiology and molecular genetics. Sanjana's parents are professors who returned to India after completing their higher education in the United States; her father earned a PhD, and her mother earned a master's degree. Inspired by her parents, she became interested in microbiology as a teenager and dreamed of working for international organizations such as the United Nations in the field of global health. She was intent on pursuing higher education in the United States, likely due to her parents' experiences. Her parents were able to help her explore options for graduate school, identifying universities to apply to. But once she was in graduate school, she had to navigate the hidden curriculum on her own, particularly the job search and immigration. The immigration system and process had worsened significantly for Indians since her parents' time. Besides, her parents had returned to India, so they didn't know about the immigration mess of the present day. When Sanjana applied to graduate school, she did not think much of the immigration system at all, but it became a significant hurdle as she explored a career path in science policy. Finding jobs in the nonprofit sector was particularly challenging for visa holders. She used lifelong learning and creative engineering approaches to find a path that aligned with her interest in global health.

Gender and Sexual Orientation

Gender and sexual orientation have an impact on the lives of international and domestic scientists alike. Male, cis-gender, and straight individuals find more acceptance and therefore have a better experience than their female, nonbinary, and LGBTQ+ colleagues. With many states passing regressive anti-trans laws and curtailing the reproductive freedom of women (post-*Roe*), many women and LGBTQ+ scientists prefer to live in liberal states, which limits career flexibility and adds to competitiveness in bicoastal scientific hubs. Add immigration barriers, and the choices become fewer for women and LGBTQ+ internationals. Harmit Mallik, professor at Fred Hutch, made an important point about the stigmas that internationals may carry: "Internationals also need to expand their belief systems and be inclusive of diverse backgrounds and identities especially if they come from conservative cultures where LGBTQ, colorism, or class-caste stigma exists. They can poison the lab environment by discriminating against peers based on those beliefs. In such cases, internationals become perpetrators instead of victims."

RECOMMENDATIONS AND ACTIONS THAT CONSIDER THE NUANCES OF INTERNATIONAL EXPERIENCE

If you read this far, you may be feeling helpless and sad. I hope you can pause for a moment and reflect on these feelings of discomfort. I listed marginalization factors to highlight that we need to take a nuanced approach to supporting international scientists that goes beyond cultural awareness, a hidden curriculum, and immigration seminars. In this section, I discuss national recommendations and some approaches to consider.

Applying Global Diversity Principles

In recent years, emerging discussions have highlighted the need to apply principles of diversity and inclusion expansively to include global diversity. NASEM's 2018 consensus study, "Graduate STEM Education for the 21st Century" (NASEM 2018) made the following recommendations.

- Correctly identify trainees' needs.
- Promote global culture by exposure to multiple worldviews.
- Consider different needs and cultural values while developing support programs.
- Develop an inclusive pedagogical framework to consider cultural difference between faculty or staff and trainee to avoid implicit bias.

Domestic DEI efforts and internationalization are traditionally treated as separate issues (for good reason). However, the current attacks on both DEI initiatives and international scholars necessitate broad solidarity and collaboration. Perhaps domestic DEI principles and global diversity and multiculturalism can be mutually reinforcing. For instance, creating a unifying space can enable learning and discussion about historical parallels: colonization, imperialism, the slave trade, the US civil rights movement, and the freedom movements of postcolonial nations globally. Such collaborations could inspire collective action to better support (and protect) domestic minorities and internationals.

Applying Intersectionality

The intersectionality framework used in learning and initiatives on DEI can also help in understanding the experiences and challenges of internationals. "Intersectionality" was coined by the legal scholar Kimberlé Crenshaw (1989) to discuss how systems of oppression overlap and create distinct experiences for Black women. The framework is applicable to people within multiple identity categories, and it can be applied to the interacting factors that create compounding disadvantages for international scientists toward informing customized support.

Framework Exercise

DEI training includes intersectionality frameworks that help you identify your relative social power and positionality based on a matrix of factors such as gender, sexual orientation, race-ethnicity, socioeconomic status, and disability, among others. You can delineate which factors contribute to higher or lower social (and professional) power for you as an individual in a group. This framework also helps you identify the interconnectedness of these factors that can compound the effect of higher or lower social and professional power.

Similarly, using the marginalization factors listed in this chapter, you can plot your relative advantages and disadvantages as they relate to each factor (e.g., passport rank). The framework will be a primer to understand your positionality in the power matrix and identify the combination of factors that drive disadvantages and unique challenges (or have compounding effects) toward finding (or providing) customized and nuanced support.

REFORMS CAN GO A LONG WAY   Approaches to supporting international scientists range from small, incremental support by individuals (e.g., advisers and managers) to larger systemic reforms. Gary McDowell's story outlines different barriers and how a single systemic reform can have a significant positive impact on the lives of international scientists. In 2011, after completing his PhD in oncology in the United Kingdom, Gary moved to the United States for a postdoc in cell and developmental biology. He did two short postdocs in Boston. Like his PhD adviser, his intention was to return home after completing the postdoc in the United States to pursue a faculty career. But life had other plans. He met his partner (now husband) during graduate school. As his partner went to medical school in Rhode Island, Gary extended his stay in the United States. Due to the two-body problem and immigration barriers (he had a J-1 visa), Gary patiently navigated unhealthy work environments. Through his own and his peers' experiences with academic toxicity, he became acutely aware of the issues in postdoc life and the critical need for reform in the scientific research enterprise focused on improving the experiences of early-career scientists. But he felt the J-1 visa prevented him from undertaking advocacy work. He remarked, "I always say this about my advocacy trajectory: I said nothing while I was on a J-1 visa, and then once I got my green card, that's when everything started to happen."

A progressive ruling on same sex-marriage changed the course of Gary's life and career. In 2013, Rhode Island legalized same-sex marriage. Gary was following the ruling closely and contemplating options. Also in 2013, the Supreme Court nullified the Defense of Marriage Act (*United States vs. Windsor*, 2013), which meant that marrying a same-sex partner in Rhode Island could be used for family-based permanent residency. On August 1, 2013, Gary and his partner were the first same-sex couple to be married in Rhode Island, and he got his green card in November. He used his career freedom to advocate for systemic changes in scientific research. He obtained a three-year grant for this work, leading him to leave academia and cofound Future of Research (FoR), a nonprofit, in 2016. He recalled, "Timing was everything. As I was a green card holder and my visa status didn't limit my career, I cofounded FoR and worked there full-time as the executive director." After about three years at FoR, Gary established and became CEO of Lightoller LLC, a scientific workforce development consultancy. The professional freedom he achieved has enabled his continued dedication to his self-professed mission: "to ensure that organizations can adapt to the changing needs of the research enterprise, focusing on the needs and issues facing upcoming generations of researchers." Gary and his husband live in the Chicago area, and he continues to exercise career leadership, recently as co-investigator on three NSF grants.

I have told other success stories, including that of Leon (Jun) Tang (see chapter 2). Jun, originally from China, applied systems thinking and strategic planning to obtain a green card during his postdoc. He did several postdocs in cancer immunotherapy research and, after years of effort, published two dozen papers and obtained a couple of patents. He took short-term financial hits through multiple postdocs but kept his eyes on the long-term dividends. Once he secured the green card, he executed his bold career visions in business of science and then entrepreneurship.

The stories of Gary and Jun and many other international scientists are inspirational. We have the benefits and privileges of scientific training and life in the United States. The breadth of opportunities in this country fuels our scientific drive and gives us more social and economic mobility than our earlier generations. Many of us do not have mountains of student debt like our American peers. If you decide to move to another country, the excellent scientific training in the United

States will set you up for long-term success. Therefore, let's be thankful for the benefits and cognizant of the disadvantages. Identifying the specific combination of marginalization factors that hinder your progress will help you seek relevant support. Finally, our diverse identities and backgrounds are a strength. Let's unite in celebrating and respecting our differences.

I end as I started, by reiterating that we internationals are not a monolith. We experience multitudes.

Supporting International Scientists

It Takes an Ecosystem

CHAPTER TWELVE

Mentoring International Scientists

Suggestions for Managers and Advisers

If you employ, supervise, or mentor international scientists, you will find yourself in situations demanding as much resilience from you as your international trainees or employees. International scientists travel far for love of science but face relentless challenges in their pursuits. They navigate a hidden curriculum at work, a restrictive immigration climate, and professional barriers with little social or cultural capital. Therefore, your first role as a mentor or manager is not to let systemic failures clip their wings and dull their brilliance. Encouragement and an optimistic outlook can help, but it can't be your primary strategy. Internationals are more likely to appreciate your honest acknowledgment of ignorance and earnest support and guidance than surface-level optimism and pathological positivity. We aren't looking for magicians but guides, collaborators, and sponsors. Besides, many of us come from cultures where directness is the norm.

Trust is the foundation of every relationship, especially those involving power differentials. Whether you employ an international scientist or guide them through their specific goals, you hold some power over and responsibility for their careers. International scientists are resourceful. However, regularly facing barriers can drive them into deficit thinking and survival mode. As a mentor or manager, help them identify tiny windows of choices in restrictive environments. They are also more vulnerable to toxic and abusive workplaces and predatory practices due to their circumstances. Encourage them to apply a growth mindset and the creativity, curiosity, and problem-solving skills they acquire in research in other aspects of their lives. In addition, advocate for reforms to improve the training and professional lives of international scientists.

In this chapter, I outline key challenges and disparities international scientists face at different professional stages, from recruitment and onboarding to work lives and advancement. I also include actionable advice and suggestions for mentors and managers on how to support international scientists at different stages (hiring, onboarding, training, and career mentoring). I highlight the imperative of advocating for international scientists and intentionally creating inclusive professional cultures. I offer insights from Harmit Malik, an Indian American evolutionary biologist and professor and associate director of Fred Hutch's Basic Science Division. Harmit has a stellar reputation in the academic community, not only for his creative and remarkable scientific career, but also for his kindness, humility, and approachability. He has a large social media following, with which he engages in honest reflections on academic life, ways to improve scientific training, and normalizing the human side of science. Many international scientists across the United States seek his guidance when facing challenges or lacking mentorship. In this chapter, Harmit kindly provided additional context from his experience and tips for faculty mentors.

HIRING OR RECRUITING INTERNATIONAL SCIENTISTS

A curious thing that international scientists encounter during a job search is the lack of preparedness on the part of the hiring team. This can take many forms; typically, it is lack of understanding and transparency about visa sponsorship. Many science-related job ads specifically require (or prefer) STEM PhDs. Yet they aren't intentional about what it takes to hire international scientists. International scientists comprise a significant portion of the scientific workforce in the United States; about 40% of PhD students and over 50% of postdocs in science and engineering are international, but many if not most job ads do not include explicit information about visa sponsorship. If you are recruiting scientists, please discuss approaches to hiring international PhDs with your recruiting team (HR, talent management, international office, or immigration attorney) in the early stages, ideally while crafting job descriptions. Understand the short-term and long-term immigration pathways; I provided a summary in chapter 7. Discuss whether a specific role is amenable to visa sponsorship (e.g., O-1 and H-1B) and how to craft inclusive job descriptions. This typically involves exploring the following questions.

- Is there a budget allocated for visa sponsorship and legal fees?
- Do the job duties meet the labor certification requirements for work visas (H-1B) as stipulated by the Department of Labor?
- Does the salary meet the minimum requirement for PhD-level scientists as required for labor certification?
- What is the minimum education requirement to hire an international STEM PhD? The minimum educational requirement of a bachelor's degree restricts the possibility of hiring international scientists with graduate degrees.
- Is there a long-term plan for retaining an international employee (e.g., a pathway for sponsoring a permanent residency petition)?

A hiring manager should discuss the possibility of an international hire early, not after the position is posted and open for applications or during interviews and negotiations. If it becomes apparent that a certain position will not include visa sponsorship for any reason, please make it explicit in the job description. As a hiring manager, you save everyone's time and energy by defining the limitations upfront. Please do not leave it to an applicant to maneuver the sticky and dreaded HR question, "Will you need sponsorship now or in the future?"

International applicants also encounter situations where hiring managers and talent recruitment or HR teams are not on the same page. A hiring manager may want an international scientist candidate, but as visa sponsorship wasn't discussed early, HR pushes back. Consequently, the candidate has to self-advocate and mediate through this situation while having little power. I have firsthand experience navigating such frustrating situations multiple times. The responsible thing to do as a hiring manager is to ask the question early and plan accordingly. If you didn't plan early but want to hire an international scientist, it's not too late. Consult with an independent immigration attorney, in addition to the company-provided attorney, to explore all options. Be patient and willing to take risks, and collaborate with the candidate and your institution's delegated immigration attorneys to put together a strong argument for a visa petition. Also prepare contingencies for delays in onboarding as getting a visa can take time even after an approved petition, especially if the candidate is overseas. Remember that the candidate or new hire is also going through this nerve-wracking process. Your calm and supportive demeanor will go a long way in allaying their fears and establishing the foundations of a healthy professional relationship.

If you (or your department or institution) commit to the idea of global diversity in graduate education, also consider inclusive processes in recruitment and admissions. International students, particularly those from LMICs, incur steep costs applying to graduate programs and the GRE due to the higher conversion rate of US dollars to their national currencies. You may want to contact your international office or graduate student or postdoc office to explore avenues for need-based financial support or application fee waivers for international graduate students from low socioeconomic backgrounds.

ONBOARDING INTERNATIONAL SCIENTISTS

Congratulations on successfully hiring an international scientist. Next, you should consider how to support their onboarding and initiation phase. International graduate students, postdocs, and employees navigate a considerable hidden curriculum in a new country, particularly if they have recently arrived. The hidden curriculum spans cultures and systems that affect their financial, professional, and personal lives. Below I highlight common systemic barriers and cultural blind spots internationals encounter in a new country and how to support them through the onboarding stage.

I. BENEFITS AND FINANCES International scientists, particularly those entering the United States for the first time, incur nontrivial setup costs. These can include expenses for air travel, advance payments on housing and rent, setup for utilities, travel and health insurance, and living expenses such as groceries and transportation. Internationals incur these costs before they receive the first month's stipend or salary. Harmit emphasized the burden of setup costs: "Many of them have already spent hundreds or maybe thousands of dollars just to go through the mechanics of applying for graduate school. It's also not a trivial thing to buy a plane ticket; for some it may be a substantial part of their salary. Besides, lots of housing properties ask for first month's rent and a security deposit upfront." Many internationals, particularly those from low-income families. take out loans to support these costs. Harmit advised faculty mentors to be informed at the very least: "Even if you can't necessarily solve all problems, you want to be aware of the challenges that your students are going through, which will ease over time as they get stipends."

You may also want to consult with central resources (e.g., graduate school, postdoc office, HR unit, or international office) in your institutions regarding need-based financial aid for the setup costs of internationals. If you don't find any avenues, perhaps you could allocate funds in your lab's operational budget to support onboarding of domestic and international members from low-income backgrounds (considering conversion of US dollars to their currencies). In the long term, though, you should advocate for equitable access across your department or school. Harmit proposed a more pragmatic workaround: "How about giving international students and postdocs a moving expense? I know that the first pushback is, 'Who's going to pay for that?' One agreeable (for institutions) option can be adjusting the setup expense from the last paycheck of the year. In other words, pre-pay (part of) their December paycheck as soon as they come in July or August." Another solution is biweekly payments instead of monthly payments across the board.

International scientists also need help with understanding salary and benefits and financial and banking systems in the United States. These can include

· opening a bank account
· building a credit history and credit score
· getting a social security card
· selecting benefits (e.g., health insurance) that comply with their visa type
· navigating federal and state tax returns

2. CULTURAL AND LIFE INTEGRATION   Most international scientists have a superficial idea of life in the United States, primarily from popular media exports. They will likely experience different degrees of culture shock and need guidance on navigating professional and daily life. Harmit elaborated on the initial culture shock many international scientists experience in the early days: "For instance, consider classroom etiquette. Many international PhD students serve as teaching assistants (TAs) during the first year of their PhD. The TA experience can be jarring for those coming from hierarchical cultures and thrown into the democratic culture of US classrooms."

Another element of surprise comes from how credit history drives access. One catch-22 for most internationals is getting a US cell phone account without a credit history. Most students rely on one or two phone companies that have flexible options for immigrants. They may also need guidance on housing and understanding lease and rent norms. Parents may want clarity on childcare options and good schools in the neighborhood. Many will also need help understanding the daily commute and modes of transportation. Is public transportation (beyond university transit) easily accessible in most neighborhoods? If not, what are the best housing and neighborhood options for them in the first year without a car. Also, what are the steps to getting a driver's license and a car? Harmit proposed a creative solution for transportation support in the onboarding months: "a monthlong Uber account, where they get a limited amount of funds for accessing Uber. This can be deducted from their paycheck in future months. It can be especially helpful in early months for those setting up life in cities that don't have great public transportation and housing costs, making it hard to live close by."

3. UNCOVERING PROFESSIONAL CULTURE AND EXPECTATIONS   Most internationals have a learning curve when it comes to understanding professional culture in the United States such as working and communication styles. Please help them understand these basic expectations.

- Collaboratively define benchmarks and milestones of success in the first quarter, semester, or year (in the graduate program or postdoc training or job).

- What is your, and the team's, preferred working style? Are you more hands-on or flexible (do you take a phased approach)? Does your team keep structured work hours, or is there a balance between in-person and remote modes of work?

- What are the norms of communication in your lab, in the department, in the field? This can take a while, but help them understand the basics. For example, I was so relieved when my graduate school advisers made it explicit how they like to be addressed. I elaborate on scientific and professional communications in the following section.

Suggestions for Effective Onboarding

For all successful relationships, a strong foundation of mutual trust, respect, and honesty is essential. You don't have to be a mentor during every aspect of their development. Instead, think of your role also as a connector: point them in the right direction. Most international offices or graduate school or HR offices should have resource guides describing the systems and processes to set up life in the United States. The National Postdoctoral Association has created a guide for onboarding international postdocs that is broadly applicable to international graduate students too (NPA International Taskforce 2023). That said, while employees have access to onboarding resources through HR or central offices, please ask your trainees questions to gauge gaps in translation and provide tailored guidance.

As managers and mentors of international scientists, you must also go the extra mile to help them understand the professional culture, environment, and expectations. To be an effective leader, strive for openness and transparency. You should collaborate with your team or department, international students, alumni, employees, and international offices to create a lab or employee handbook that communicates the expectations of all group members. You can add a supplemental chapter on resources and know-how for international members of your group. Harmit added, "People come from different backgrounds and have different value systems. You should discuss the values of the lab. It's important to talk about values because one may not come in with that mind-set but can be trained."

I also repeat the same to you as I did to your international mentee or employee in previous chapters. Have the courage and humility to acknowledge that you don't know what you don't know. You can lead by connecting them to other mentors with direct knowledge and experience of their specific developmental needs by encouraging them to cultivate a team of mentors, including near-peer (alumni) and peer mentors. Many affinity-based international student or employee groups may have a peer mentor program, or you could create a near-peer buddy system in your lab or department.

TRAINING INTERNATIONAL SCIENTISTS

Education systems vary globally. Many do not follow the strictly liberal arts framework typical in the US undergraduate curriculum. Therefore, as mentors you will need to understand international trainees' prior knowledge and frame of reference. While they will have a fundamental understanding of the "scientific method," they may be at different levels of fluency when it comes to the process of scientific research, from research design to data collection, analysis, and communication. It's understandable to take a hands-on approach at first, but be mindful not to veer into a perennial micromanagement mode. That will hinder their creativity, sense of agency, self-efficacy, and confidence.

International trainees (and employees) may also need mentoring with norms of science communication in both broad and field-specific manners. This includes typical standards of oral and poster presentations, scientific manuscript preparation and grant proposals, peer review (and responding to editors), and effective communication for nontechnical audiences. Also, if you observe that students don't ask questions regularly in individual or group meetings, you will need to create a more welcoming and psychologically safe environment for them. Encourage them to contribute by asking their opinion specifically. Please be patient and thoughtful as many international trainees are non-native English speakers. They often translate their thoughts from a different language to English, orally and in the written form. Please also ensure that your feedback doesn't come across as intimidating or, worse, condescending. In the fast-paced world of scientific research, it may be frustrating, but put yourself in the shoes of your trainees, who regularly navigate unknown terrain and a hidden curriculum with grace. Internationals are fast learners and resourceful. They can learn quickly if you show them a few good examples and why those examples work well. It is also crucial that you affirm or compliment their accomplishments and progress. Your approachability and kindness will pay off in the long run. Confidence in one's abilities is just as important for success as skill and talents. As your approach to mentoring can boost their confidence or destroy it, please put thought into tailoring your approach.

As an adviser or supervisor, you hold a position of power and influence over your trainees' career and well-being. This power differential can have existential implications for international trainees as they need your support for maintaining their visa status in order to stay in the United States. Academic productivity (e.g., high-impact papers, patents and grants, strong letters of recommendation) are not just important for career advancement, but are tied to immigration advancement and permanent residency. Keep that in mind when assigning projects to internationals or helping them select projects. Strike a balance between long-term risky projects and applied projects with a likelihood of faster publications. Also, be intentional when discussing authorship on collaborative projects. Evidence of independent scholarship and a significant contribution in one's field are important for their immigration path too. Therefore, enable ways for them to highlight key contributions in advancing the scientific field. For example, make their contributions visible in media highlights of published work, perhaps even include their interview. Regular discussions to define short- and long-term goals are important as international scientists need to justify extending their student or work visas. In chapter 6, I described how to develop SMART goals and individual development plans (IDPs) for the benefit of your international trainees. You should also foster a culture that balances serendipity and structure and encourages accountability. Craft an annual IDP in discussion with your international trainee(s), which includes project and training goals, skills development, and career goals. Online IDPs such as myIDP by AAAS and Chem IDP by ACS are good starting tools. Also, regularly check in about their immigration goals and needs. You don't need to probe for details about their immigration situation; simply offer support. Ask if you can help in any way; a small gesture reflects care and openness. It is essential that your actions, not simply words, show care too. Maintaining one's visa status involves a lot of paperwork and dealing with bureaucracy. Some managers treat it like it is not their problem. Be a responsible and inclusive mentor and manager by treating immigration requirements as work that can be done during working hours.

THE HIDDEN CURRICULUM IN NETWORKING AND COMMUNITY BUILDING

Among the best advice to early-career scientists is to build professional networks for career and professional development. Scientific conferences are fertile grounds for global knowledge, networks, and collaborative potential. That said, academic conferences can feel overwhelming and exclusionary to international scientists.

Uncovering the Hidden Curriculum of Networking and Conferences

As an adviser, you should clarify the professional norms of scientific talks and posters in your field or specific conferences for your international trainees and employees. These include how to write conference proposals and abstracts, how to give effective talks or posters, and how to navigate audience Q&A. While many labs practice talks with peer feedback, navigating Q&A and professional introductions and pitches is not usually discussed. Collaborate with science communication experts to host workshops in the department. Also, collaborate with career and professional development experts in your institutions to include workshops on professional networking, for instance, how to introduce oneself and one's research in one or two minutes in scenarios like sharing the elevator or sharing space before a talk or standing in the lunch line with an eminent scientist.

Next, let's discuss social situations. As a faculty member or seasoned scientist, you have more social power in these scenarios to make international scientists feel included. If you share a lunch table with graduate students and postdocs, include them by asking questions about their research. It's hard for all students, particularly internationals and first-gen students, to confidently interject their thoughts in a large group. While I am on the topic of conversations at conferences, I offer a quick note to journal editors who may be reading this. While your job is looking for the best research ideas and engaging with senior scientists and faculty, please include early-career scientists and international scientists.

Conferences have a way of compounding imposter syndrome, so please be mindful about how your actions will affect the next generation. The overuse of popular culture references and small talk actively disadvantages international scientists. If you must engage in popular culture references, you can politely ask any international student in the crowd if they understand them and provide a brief context if they don't. Early in my graduate school life, I was weary of conference networking as it often veered into pop culture references to sports, media, and local geography. Over time, I realized a lot of community and potential collaborations are forged during meals, receptions, and optional group activities or at bars. So I started participating, nodding along at first, listening, slowly asking questions, drawing parallels with my life in India. My favorite parts were discussions among international scientists about similarities and differences in our cultures and habits. I faced my fears, but not all internationals will. Therefore, it is prudent to create an inclusive conference culture for scientists from diverse backgrounds.

Finally, if you are a conference or event organizer, include international scientists, particularly early-career ones, in the planning committee. They can provide key inputs regarding how to make the conference or event inclusive. Social community building events can be counterproductive if they focus on US-centric topics and trivia. A critical systemic aspect of organizing inclusive conferences is picking venues that are inclusive for international graduate students and postdocs. For locations outside the United States, international trainees, even those at US universities, need to apply for visas, which may take time and money, so allot plenty of time and travel awards for international scientists.

FOSTERING A GROWTH MINDSET AND CREATIVITY AMID RESTRICTIONS

Encouraging curiosity and creativity should be at the core of your mentorship approach for all trainees, particularly international trainees. As international scientists regularly encounter systemic barriers and challenges, they may fall into the habit of a deficit mindset. The most onerous form involves structuring one's life within the confines of immigration feasibility. As a result, international trainees may limit use of their imagination and operate in a comfortable or pragmatic zone. Therefore, as a mentor, make an extra effort to encourage them to embrace a growth mindset and creativity in their career goals. In this section, I offer suggestions on how to support international scientists in harnessing creativity for success. You can also refer to chapter 3, where I outline approaches for creative engineering through systemic barriers by proposing a framework.

First, early in the relationship, you must learn about the international student's mindset, motivations, strengths, and areas of growth. In parallel, learn about the systemic barriers that international scientists face. With some background understanding of the challenges, you can collaborate with international scientists to craft a version of the design framework in chapter 3. Specifically, you can collaborate with your international mentees on defining success in tangible ways, clarifying their purpose and goals, and connecting them to people and resources. As I mentioned previously, in addition to an adviser or mentor, you can serve as a champion and sponsor. Don't assume the burden alone; crowdsource when you can. Leverage connections and institutional resources to explore options and be creative partners in the process. Let me clarify via two examples. In chapter 3, I suggested to international trainees how to navigate two systemic barriers (access to fewer fellowships and restricted career paths due to immigration) to map paths to success using creativity. I use the same test cases (fellowships and career) to outline how you, as mentors and managers, can be creative partners.

Fellowships

International scientists are not eligible to apply for most federal research fellowships. First, understand their primary motivations for obtaining a fellowship or grant (e.g., for career or immigration or both). Second, explore diverse avenues. For instance, they may be unaware of corporate sector or foundation fellowships that are open to internationals. To advance their goals, you can help them with funding discovery, in addition to providing training in grant writing. Most research-intensive universities use funding discovery tools (e.g., Pivot) that list fellowships open to international students and postdocs. Moreover, Johns Hopkins University developed an open-access fellowship discovery database for graduate students and postdocs. Also, connect them with staff in your university's research development or corporate and foundation relations offices to explore funding options for international scientists. You can mobilize your network to identify international scientists who have secured funding and point your mentees to that direction. If all else fails, be an enabler by opening opportunities. Explore opportunities for them to be a coinvestigator with you on a research grant. Your international scientist mentee or employee may be scared to ask; bring it up as an option. This will assure them you will do everything to support them.

Immigration

Long-term career outcomes studies of STEM PhDs (e.g., Coalition for Next Generation Life Sciences, PhD Career Pathways by Council of Graduate Schools) indicate that most STEM PhDs have dynamic, nonlinear careers. Diverse careers are inherent to scientific training to the extent that PhD programs in STEM now take pride that scientific training enables workforce and economic development across sectors. However, career advancement is an intense struggle for international scientists because the employment-based immigration paths in the United States have not caught up with workforce realities. If anything, the political climate after 2016 and recent layoffs in the tech sector have made employers in many fields risk averse. As a result, international scientists face hurdles in securing (and keeping) jobs, particularly those not closely related to their research, within and beyond academia. To support them, first learn about the immigration paths and advancement hurdles. As I mentioned earlier, regularly collaborate with international offices and immigration attorneys in your institution or organization to assess short-term and long-term solutions. Second, connect them with alumni and scientists in diverse careers, particularly other international scientists. Third, encourage them to pursue professional development beyond the lab. Encourage their participation in career panels and workshops, industry research collaborations, and internships. You can advocate that your institution make internships accessible to all international students. Your department or school can leverage the curricular practical training (CPT) in collaboration with the international office and create policies to make internships accessible to international students. To ensure compliance with visa requirements, you could create a graduate-level course on internship or help them include lessons from experiential learning in their dissertation.

Omanjana Goswami's faculty adviser supported including her science policy internship as a dissertation chapter, which ultimately helped in her career journey. Omanjana moved to the United States from India as a PhD student in environmental sciences. A lunch panel on science policy at a conference piqued her interest. She recalled, "Like many graduate students, I attended the panel for its three-course menu, but that free lunch changed the course of my life." She preferred solving applied problems, and the science policy field provided the opportunity to pursue community-engaged research and policy. She investigated science policy internships and fellowships open to international students and came across many scientific society fellowships, including the National Oceanic and Atmospheric Administration (NOAA) Knauss science policy fellowship. With her adviser's support, she applied and leveraged CPT to participate in the NOAA science policy fellowship. She stayed enrolled in her doctoral program, doing research credits as she had completed all her course requirements. Rutgers University's international office did a fantastic job keeping her on track to maintain her F-1 student visa status by enrolling her for a single credit course tied to the science policy fellowship, and including it as an auxiliary chapter in her dissertation helped. The latter showed a direct connection between her PhD research and policy, making it easier to use immigration paths like OPT (particularly STEM OPT) and an H-1B visa to transition into the science policy field. Currently, Omanjana is a scientist at the Union of Concerned Scientists, a nonprofit research organization. With her adviser's support, Omanjana found the confidence to be more adventurous with her career.

I can't emphasize enough that your allyship and support are crucial to convincing your international trainees that they have an element of choice amid very restrictive systems. Your openness to supporting their aspirations will go a long way in building their confidence and exercising creative imagination. You can help their job search, without judgment. If they decide to continue in academia, offer constructive comments on their applications and write letters of recommendations. Serve as a sponsor by connecting them to other experts in the field at conferences or sponsoring them for professional development opportunities. Celebrate their accomplishments loudly on social media. If they decide to pursue a career outside academia, help them contact alumni or connections you have in other sectors; encourage them to meet with career and professional development educators at your institutions. If they want to explore global careers, connect them to your international networks and help them identify reputable international institutions. Finally, have regular conversations with them on realistic timelines for completing their research work with you. This helps them plan their visa extensions or transitions accordingly.

Harmit offered this advice for mentors about supporting trainees interested in nonacademic careers: "A faculty adviser's job is to train scientists well and support trainees irrespective of their career goals. Many faculty fall into the trap of thinking that if a person doesn't aspire to an academic career, they're going to mentally check out. But that can also happen when someone gets an academic offer as they're in the lame duck phase. You can't talk someone out of their aspirations. Just offer your support and help them find the best path. You may not be the best mentor for their new career, but you can help them find connections. And in the meantime they can still do excellent work."

SUPPORTING THE WELL-BEING AND DIVERSE STRENGTHS OF INTERNATIONAL SCIENTISTS

International scientists shoulder a lot of responsibilities with little social capital in a new environment. They face steep odds to succeed away from their families and loved ones. Homesickness, the hidden curriculum, the uncertainty of research, fewer career opportunities, and immigration restrictions can negatively affect their mental health. To make matters worse, they are easy targets of workplace abuse. There are many stories of managers overworking or stalling the promotion of international scientists by using immigration as an excuse. Also, bad behaviors and inequitable systems disproportionately affect nonwhite immigrants and those from the Global South or those with less socioeconomic privilege. I provide a more nuanced explanation of mental health challenges and the overlap of DEI principles in the immigrant experience in chapters 10 and 11, respectively. As someone with more relative power and influence over your immigrant trainees and employees, I urge you to examine ethics and equity in your work environment and team dynamics. Be thoughtful about team dynamics in your group. Ensure equitable access to knowledge, mentorship, opportunities, connections, and support among your team members. Within and beyond your team, if you see something, say something. If you know colleagues who resort to toxic and problematic behaviors, report them via the proper channels. Ghost mentoring may seem like an easier solution, but it is not sustainable; you will likely find yourself mentoring many in such precarious positions. This burden also falls disproportionately on women and minoritized faculty and managers. Besides, the cycle of abuse persists when perpetrators aren't called out. Therefore, by not speaking up, you risk collective dehumanization, leading to attrition and exclusion of generations of talented scientists (mentors and mentees). Harmit offered a slightly different recourse. While he agreed that abusers go unchecked and thrive without punitive measures, he prefers to focus on the needs of students and more achievable solutions. He said:

> The biggest issue is that immigration status is directly tied to traineeship, making trainees vulnerable. Supervisors who abuse this privilege get away with it because of their positional hold over trainees. Whenever I get called into such situations, I prioritize the best interest of students. Many faculty disproportionately spend time ghost mentoring, and they need to be rewarded. I am a big fan of solutions ensuring the student is no longer in an abusive environment, helping them transition to a better situation and rewarding ghost mentors. That said, immigration offices can take punitive measures such as rescinding the abusers' ability to hire immigrant scientists for a period of years, which will directly affect their ability to do good science.

I echo Harmit's suggestion that you can reach out to Ombudsman offices, colleagues in graduate school, postdoc offices, or HR for help and guidance. If you are a departmental director or chair, create avenues for reporting without repercussions. Have accountability in place, such as hiring freezes if multiple accusations are reported. Many such policies exist but are often not used when it comes to mistreatment and abuse of international scientists.

It is a delicate balance to guide someone regarding professional norms while not stripping them of their unique strengths, agency, and identity. Internationals regularly struggle with losing their identity and voice against normative expectations. Expectations of "professionalism" and "professional expression" present the most direct example of this problem. Internationals are expected to talk or write (or worse, think) like a "proper scientist," which typically means like a white, male scientist. As an adviser or mentor, you have the best intentions when it comes to training the next generation for success. But what skill sets, tools, mindsets, or approaches do you prioritize to achieve that goal? Do you rely on mentoring shortcuts while navigating the race for scientific productivity? Are you creating versions of yourself (or your mentors) to shape your legacy? Are you expansive in your mentoring to provide space for diverse thinking, perspectives, and experiences to flourish without judgment? It takes humility to learn from the diversity of thought and experiences that generations of international scientists bring from different parts of the world. The most successful mentors empower the varied, imaginative, and innovative mindsets of their diverse team members, amplify diverse strengths, and actively block deficit thinking—in themselves, as much as their teams. You can fully realize the potential of your multicultural team by striking this key balance of guiding and learning as a mentor.

MODEL GOOD BEHAVIOR

Harmit astutely pointed out that ultimately modeling good behavior is the best thing any mentor can do to ensure that their trainees will be good mentors in the future. He summarized how four mentors had influenced his path and shaped his mentorship style.

> A mentor's job is to model to their mentees what a "successful" life and career can be. For instance, if you model work-life balance and celebrate small victories in the lab, that's what trainees will emulate. I emulate models from four significant mentors. My first mentor was a professor at the Indian Institute of Technology (IIT), Bombay. I was a chemical engineering student, and he was a biosciences professor. He taught me introductory molecular biology in one-on- one sessions. I will always remember that this busy and overworked person spent hundreds of hours teaching me (someone he wasn't responsible for). Consequently, I carry a debt of

gratitude forward. I learned from him that I never want to instantly turn away someone "like me" because they didn't fit stereotypes of a successful student. Whether they thrive or not, I can give them thirty minutes of my time on career guidance and psychosocial or well-being support. My PhD adviser was my next mentor. He modeled rigor and scientific excellence but also work-life balance. He prioritized family time; very few things would keep him from dinner with his family every day. He was the first one in the lab at 7:30 a.m. but left before 5:30 p.m. When you train in that environment, you think it's normal to have work-life balance. He never required that we arrive at a certain time or work until a certain time, but he did require that we were invested in our project. From him, I learned to give students the benefit of the doubt until they have given you reasons not to. My postdoc adviser was my third mentor. He taught me that you can do the most creative things if you know that your adviser is in your corner and has your back. He encouraged us to propose and try out risky ideas. It's an empowering thing. You can make students live up to your expectations by just reassuring them that you have their back. After I started my lab, I built a beautiful relationship with my senior colleague, Mike Emerman, for the last twenty-one years. He's a virologist and I'm an evolutionary biologist, almost perfect symbiosis. We are very different people. We write differently, and if you give us an idea, we'll probably have completely different thoughts about what to do next or how to interpret it. But we have immense, unshakable respect for each other. He is also a fantastic mentor, more structured than me. But I try to adapt what I can from his mentorship style.

Model the behavior that empowers your mentees. Mentoring international scientists can be frustrating as you will face systemic roadblocks with them. The systems that affect international scientists, primarily the immigration system, in the United States are dated and complicated. You will feel powerless and struggle to make peace with things that are beyond your control, but do not let that stop you from advocating and lobbying for systemic changes at the institutional and national levels. International scientists face many structural barriers due to federal policies that are beyond an individual's control. As a manager or mentor, you may feel frustrated that you cannot offer adequate short-term solutions. But you can advocate for long-term reforms.

You can mobilize colleagues to advocate for institutional resources such as fellowships that are open to international trainees, collaborate with your institution's offices of federal or government relations to lobby for immigration reforms, and join forces with national organizations such as AAAS, NSF, the Federation of American Scientists, the American Association of Universities (AAU), the Council of Graduate Schools, the National Postdoctoral Association (NPA), and many other scientific societies to advocate for inclusive systems, whether its opening access to federal fellowship opportunities or improving the immigration quagmire. International scientists deserve it, and the future of the US scientific enterprise depends on it. Remember, your initiative is enough to inspire your employee or mentee. International scientists do most of the legwork in mapping immigration pathways. All they need is to know that their mentors are in their corner, that they are not alone, and that they belong. See their humanity, beyond their scientific excellence. That is the best gift you can give your international colleague.

KEY HIGHLIGHTS

- Develop hiring practices that intentionally include international scientists.
- Learn about the hidden curriculum to effectively onboard and train international scientists.
- Understand the barriers to success for international trainees to creatively design short-term solutions and advocate for reforms. Use your position to sponsor them for opportunities that enable success.
- Become allies and mentors to support the diverse career aspirations of international scientists. Mentor when you can; connect them to people and resources when you can't.
- Collaborate with institutional offices to leverage all capabilities and support infrastructure.
- Model good behavior to set examples for mentorship among the next generation.

CHAPTER THIRTEEN

Systemic Reforms to Attract and Retain Global Scientific Talent

WITH ADRIANA BANKSTON, PHD

International scientists (foreign born and temporary visa holders) are key drivers of scientific research and innovations in the United States, maintaining the nation's competitiveness in science and technology (Bernstein et al. 2022). However, specific factors within the scientific enterprise, a dated immigration system, insufficient stipend and wages to cover the cost of living, fewer independent funding avenues, inadequate mentorship and career support, and hostile work environments deter talent from staying and thriving in research environments in our universities and national laboratories. These issues have been highlighted throughout this book and documented in several publications and reports, including a well-known paper on rescuing US biomedical research from its systemic flaws (Alberts et al. 2014). In previous chapters, I outlined approaches for individuals and international scientists and their mentors and managers to creatively navigate restrictive environments to achieve professional success.

That said, long-term reforms and improvements require systemic change. Several organizations are tackling this problem on the systems level, including universities, national associations (e.g., AAU, AAAS,

NAFSA, and NPA), often advocating for more welcoming environments for international scientists. In this chapter, we further clarify structural barriers and recommend policies and approaches for supporting international scientists in the United States at an organizational and systems level. Building from chapter 1, we elaborate on challenges faced by international scientists and offer actionable steps and recommendations. We also highlight key national initiatives advocating for international scientists and add our voices to advocate for policy reforms.

INTERNATIONAL SCIENTISTS PLAY A KEY ROLE IN STEM WORKFORCE DEVELOPMENT IN THE UNITED STATES

The increase in international STEM PhDs in the United States has important workforce implications for the science and engineering fields. According to 2022 data from the American Immigration Council, workers with STEM training and skills are the most sought after in the labor market, and foreign-born workers make up a growing share of the country's STEM workforce (American Immigration Council 2022). For example, as of 2019, immigrants made up 23.1% of all STEM workers in the entire country, which increased from 16.4% in 2000. This highlights the growing importance of immigrant workers in STEM occupations in every state in the country and across different STEM fields. As the US economy continues to become more technology dependent, more qualified STEM workers will be needed by employers to fill vacancies and maintain US competitiveness. While US universities still lead globally in training international STEM students, continued investments are needed to retain these international scientists in research and development and advancements in critical technology and engineering fields such as AI, cleantech, and bioengineering in a globally competitive market.

INTERNATIONAL SCIENTISTS FACE CHALLENGES IN TRAINING AND RESEARCH INDEPENDENCE

Multiple factors contribute to specific challenges in the training experience of international graduate students and postdocs that may affect academic retention and success in independent research careers. First, like first-generation domestic students, international graduate students and postdocs navigate the hidden curriculum, or unknown rules, of academic training. In the absence of established professional networks and cultural know-how, they primarily rely on guidance from their faculty advisers and institutional support to understand the academic environment, training goals, milestones, and tools required for professional success. For example, many international graduate students report challenges due to an inadequate understanding of the format and expectations of the comprehensive exams required for doctoral candidacy. The examination for candidacy is particularly important for retention in doctoral education and overall academic career interest, especially when interest in academic careers among current STEM graduate students is declining. A 2017 study provided two key insights (Roach and Sauermann 2017). It found that although the majority of students begin PhD training with an interest in academic research careers, over time 25% lose interest entirely. And this decline in interest was partly due to the misalignment between students' changing preferences for specific job attributes and the nature of the academic research career itself. Interestingly, the change in career interests is often seen after the candidacy exams and may correlate with their experience. Therefore, thoughtful mentorship by faculty advisers and a culture of transparency and care in research labs are particularly important for the success of international scientists. A helpful resource is the Science of Effective Mentorship in STEMM initiative by the National Academies of Science, Engineering, and Medicine (NASEM), which provides recommendations for faculty advisers and departments to create and support viable and sustainable mentoring support systems (NASEM 2019).

Second, as discussed in chapter 1, international scientists have access to fewer fellowships and grants for exploring independent research ideas. They typically rely on faculty-directed or university-administered assistantships for training. Reliance on institutional or faculty-directed funding can also result in a lack of agency among international scientists to pursue independent research ideas, define the finish line for training, and explore professional and career development opportunities such as internships. As Gertrude's story highlighted, international scientists are at higher risk of unexpected terminations when faculty members lose funding. All these factors can contribute to a decline in mental health and well-being. Moreover, this leaky pipeline to the professoriate has an impact on global diversity in academic careers. As noted in chapter 1, in 2022, only 21% of international PhD graduates from US universities committed to academic careers. Therefore, we need to create more opportunities for—and awareness of—diverse, independent funding streams for international scientists if the academic sector wants to retain this global scientific talent pool of PhDs trained in the United States.

Improving the Research and Training Experience of International Scientists

Given the constraints, transparency and supportive environments are critical. It is important to paint a clear picture of the basic realities of academic training and pursuing a career in academic research for international graduate students and postdocs. You may be wondering how we approach this goal. The short answer is that it will require mobilizing an ecosystem of support—both national and local—as well as systemic changes. For starters, in two separate consensus study reports—*Graduate STEM Education for the 21st Century* (NASEM 2018) and *The Postdoctoral Experience Revisited* (NASEM 2014), NASEM assessed graduate and postdoc training in the United States and provided principles, action points, and recommendations for enhancements. These findings and recommendations can be used to inform internal and external stakeholders to act on these issues. Stakeholders include faculty advisers, university administrators, and professional organizations, as well as policy makers who can introduce policies to improve STEM training and make the environment more welcoming. While these efforts are useful for overall reforms that can lead to systemic changes over the long term, the current reality is that many international scientists are unable to thrive in the US research enterprise due to the lack of funding mechanisms supporting them. This includes federal grants that place eligibility restrictions on non-US citizens. In the short term, recommendations to include more international students and postdocs in institutional training grants may open some avenues (Roach and Sauermann 2017). Nevertheless, there is still a need to change policies that prohibit the funding of non-US citizens on fellowship and training grants, which could come from legislative changes or executive branch actions and programs from the federal government. Further, more transparency, access, and support for diverse streams of funding (e.g., corporate and foundation sources) for international scientists can level the playing field for competitiveness. In its 2024 consensus study report called International *Talent Programs in the Changing Global Environment*, NASEM made a similar recommendation with regard to transparency. The report recommends creating a public-facing dashboard that showcases important metrics on international talent, including funding opportunities for noncitizens from federal and state governments, industry, and philanthropy (NASEM 2024). Collective action can help make the STEM workforce more aligned with how the population looks and elevate the contributions of foreign-born scientists to the future of science in the United States.

Impact of Lack of Career Diversity among International Scientists

Chapter 1 showed that most international scientists are employed in academia or industry, most likely in research roles due to visa and immigration constraints. They are vastly underrepresented in nonresearch careers in academia and industry (e.g., management) and in the nonprofit and government sectors. The stories of individual international scientists woven throughout the book highlight the ingenuity and sheer willpower needed to break into these nontraditional roles. As mentioned in chapter 1, less than 3% of international PhDs are hired in the nonprofit sector and less than 2% in the government sector. Such systemic exclusion of international scientists has repercussions. Systemic exclusion of international PhDs, particularly those trained in the United States, from the government sector has opportunity costs too. International scientists can apply their scientific knowledge and degrees effectively as well as fill crucial gaps in the scientific workforce by entering the federal workforce, for example, in science policy. A report from the Union of Concerned Scientists discusses the shortage of scientific expertise in the federal agency workforce from 2016 to 2020, partly due to a science-unfriendly administration (Desikan et al. 2023). Filling government positions with scientific experts is critical for our nation's economic viability and prosperity and for the ability of government decisions to be made in an evidence-based manner where scientists have a seat at the table and can influence important national priorities leading to improving national security. This loss of federal scientific capacity affects not only government decisions but also the quality of scientific work and the ability of a given government body to function with the same efficiency. As we detail in the last section, the federal government under President Joe Biden was starting, albeit late, to meet the moment by taking policy-based actions to achieve reforms. The Trump administration is nullifying gains and setting us back with reductions in the overall federal workforce, including scientific talent.

THE ROLE OF UNIVERSITIES IN SUPPORTING INTERNATIONAL SCIENTISTS

Universities must take leadership in supporting their international scientists through data transparency, professional development support, and access to diverse networks. First, beyond reporting on the career outcomes of their PhD graduates and postdocs, universities should actively engage with their international PhD and postdoc alumni. Alumni play a key role in diversifying networks, mentorship, and access to opportunities for international trainees. Second, PhD and postdoc training in universities should be comprehensive and robust, providing dynamic skills to engage with wicked problems and knowledge of the employment spectrum for science and engineering trainees, agnostic of their desired career outcome. For example, in the field of AI, which is integrated into many sectors of our society, workers need to build new literacy skills and adapt to the needs of the field. While career and professional development programs for PhDs are growing, most of them operate on the university-wide level and are cocurricular in nature. This presents barriers to participation for all trainees, particularly vulnerable populations such as international trainees who cannot assertively and openly get out of labs to participate in career and professional development initiatives. As outlined in the AAU's PhD Education Initiative (AAU 2019), foundational career and professional development should be embedded in graduate programs or postdoc training within departments, signaling endorsement by faculty and an inclusive departmental culture. Third, universities must provide international graduate students with equitable access to internships and nonacademic experience. As a growing number of science and engineering graduate programs entertain the idea of site visits, internships, and corporate research fellowships, they must consider ways to make them available to international PhD students. At the very least, departments can learn about different ways to make CPT available for off-campus internship opportunities for all international students with the help of the international students and scholars' offices.

THE ROLE OF GOVERNMENT AND ADVOCACY ORGANIZATIONS IN ADVANCING AND SUPPORTING INTERNATIONAL SCIENTIFIC TALENT

The federal government can lead in creating a positive system and environment for scientists. Systemic barriers to the inclusion of foreign-born scientists persist and are often amplified, leading to a lack of sense of belonging. Moreover, explicit signaling from the federal government that immigrants are not welcome in our scientific enterprise, especially since 2016 and through the COVID-19 pandemic, has worsened the situation. For example, in 2020, President Trump signed an executive order titled, "Proclamation Suspending Entry of Aliens Who Present a Risk to the U.S. Labor Market Following the Coronavirus Outbreak," which followed a series of additional executive orders targeting the foreign-born workforce, deterring foreign talent from our research enterprise (White House 2020). Recent survey data published in 2022 tracked the self-reported impact of the COVID-19 pandemic and the Trump administration's executive actions on the early career experience and perceived university responses to combat these detrimental impacts. The survey showed that foreign-born early-career researchers in the United States experienced a significant negative impact on their mental health, academic life, flexibility to leave and enter the United States, and sense of belonging during and following the pandemic (Jorgensen et al. 2022). The indiscriminate visa revocations of international students and large reductions of federal investments in academic research by the second Trump administration will worsen recruitment and retention of international scientists in the future (Blake 2025; *Inside Higher Ed* 2025). In fact, 75 percent of the approximately 1,600 scientists who responded to a 2025 *Nature* poll said they are considering leaving the United States following the recent disruptions to federal scientific funding, with Europe and Canada being top choices for relocation (Witze 2025).

Several scientific and advocacy organizations have responded to prior government actions, including Future of Research, cofounded by Gary McDowell, which has long championed the need for changes in response to federal government actions. The Federation of American Scientists has also worked to strengthen the US immigration system to ensure that scientific experts and entrepreneurs with innovative ideas can build lives in the United States (Milliken and Santiago 2021).

The Biden administration took remedial measures to support foreign-born scientists and their contributions to US research and development. Federal agencies and their most influential boards and committees highlighted the importance of international STEM talent for our nation's research enterprise, including in a fact sheet from the National Science Board (National Science Board 2022) putting forth mechanisms to accomplish two goals:

- **Proactively and broadly attract talent** through more sources of stable, long-term funding in academic labs and support for founding and growing companies, calling for the United States to continue to champion core values in science and engineering such as openness, fairness, and accountability.

- **Reduce deterrents to studying or working in the United States** by designing a streamlined and expanded visa system and research integrity policies that reduce the administrative burden and maintain a welcoming atmosphere for international students and workers.

The NIH Advisory Committee to the Director (ACD) recently implemented a working group to reenvision postdoctoral training. After numerous listening sessions and responses from all stakeholders, including postdocs, ACD made recommendations that include a sizable increase in salaries of postdocs (~$70,000 annually), dedicated time for professional development, and a cap on the length of postdoc positions toward securing the future of academia's research workforce (NIH 2024).

In terms of immigration reform, legislators on Capitol Hill attempted to make certain STEM advanced degree holders eligible for permanent resident status. This will allow international PhDs to remain in the United States following their graduation and remove barriers to working in this country. A notable piece of legislation in this regard is the bipartisan Keep STEM Talent Act, to expand the ability of America's STEM workforce to compete in the global economy and to provide research and educational opportunities for international students in STEM disciplines (Foster 2023).

Much of the immigration reforms needed for foreign-born scientists also depend on the number of temporary work (H-1B) visas for foreign-born, high-skilled workers employed by US firms. The number of H-1Bs offered has increased substantially since 2010, which can help retain more foreign-born scientists in our research enterprise (NSF 2019). However, much of this increase is specific to fields and certain job transitions, thereby limiting flexibility. Therefore, the Department of Homeland Security released a Notice of Proposed Rulemaking to reform the H-1B visa rules, which would allow these visa holders to easily transition into new jobs, have more predictability and certainty in the renewal process, and have more flexibility or better opportunities to apply their skills (Kaushik 2023). The philanthropic sector is using innovative approaches to encourage looking beyond H-1B. For instance, Talent Mobility Fund, a program of Renaissance Philanthropy, deploys philanthropic resources to increase uptake of existing but underused lawful immigration pathways (such as O-1 and J-1) for science and engineering talent.

The long and convoluted path to permanent residency, particularly for Indian- and Chinese-born scientists, is driving international scientists toward global talent programs that provide easier paths to permanent residency and citizenship. NASEM's 2024 consensus study report, *International Talent Programs in the Changing Global Environment*, recommends creating easily navigable pathways to permanent residency and citizenship for foreign-born STEM talent, with explicit eligibility for those graduating from US universities (NASEM 2024). The report also recommends authorizing additional green cards (which are exempt from country caps) for international STEM talent working in critical areas of science, technology, and engineering. Further recommendations include calls for the executive branch to maximize use of its authority to retain foreign-born STEM experts already in the United States by updating labor certification and easier access to permanent residency.

Overall, these efforts can also help improve the labor market where more foreign-born scientists are included. Through the CHIPS and Science Act, the US government invested $280 billion in efforts to bolster US semiconductor capacity, catalyze R&D, and create regional high-tech hubs and a bigger, more inclusive STEM workforce (Badlam et al. 2024). While the Biden administration initiated and implemented many progressive STEM R&D policies and infrastructures, the second Trump administration is in the process of suspending or discontinuing those efforts.

Immigration policy discussions have been at the forefront in Washington for the past several years. For example, the Bipartisan Policy Center has examined the role of immigrants in the US economy and reforming immigration policy where bipartisan agreement is possible (Bipartisan Policy Center 2022; Butler 2023). While the CHIPS and Science Act presents tremendous opportunities for economic development in the United States, we won't realize the impact of this legislation if we don't have enough highly skilled workers who sustain both regional growth and international competitiveness. This starts at the student level. Recommendations for reforms to student visa and high-skilled immigration policies include ensuring that international students and scholars can come to the United States to study and work and have a clear path to a green card (Conrad 2022). The US Department of State's Bureau of Educational and Cultural Affairs (ECA) issued an "Early Career STEM Research Initiative" (DOS 2021) and guidance that will facilitate additional academic training for undergraduate and graduate students in STEM fields on the J-1 visa for periods of up to 36 months. In addition, the Department of Homeland Security included 22 new fields of study in the STEM OPT program in 2022 through the Student and Exchange Visitor Program (SEVP) and updated its policy manual related to "extraordinary ability" (O-1A) nonimmigrant status regarding what evidence may satisfy the O-1A evidentiary criteria. In this 2022 update, DHS clarifies how it determines eligibility for immigrants of extraordinary abilities, such as STEM PhD holders. These changes pave the way to retain foreign-born talent in US research by providing additional opportunities.

We are at an interesting time when scientific and technological advancements are moving faster than ever. Given the growth of the population, international trainees and scientists are at the center of research-based discovery and innovation. To maintain its scientific eminence in an increasingly multipolar world, the United States must create incentives to retain international scientists, particularly those trained in US universities. Toward that goal, the academic community—students, postdocs, alumni, faculty, staff—as well as allies (professional societies, advocacy groups, funding agencies) and the government are collectively responsible for creating an enriching ecosystem where international scientists can thrive, solve wicked problems, and make meaningful societal impacts. International scientists are the powerhouse of scientific discovery and technological inventions and advancements. If the United States wants to continue to attract and retain global scientific talent, systemic improvements are prudent.

The Paths We Travel

Celebrating International Scientists

We international scientists invest a great deal of time and energy in scientific pursuits and carving careers. We seldom if ever pause to reflect on how far we have come. We move from goal to goal, relentlessly striving for excellence and determined to overcome adversities. Let's stop and celebrate ourselves and the international community. Let's take pride in our journey. I hope the stories of international scientists in this book showcase our ingenuity, our diverse modes of leadership, and our sense of community. I urge you to share your stories too. You may become a role model for another international scientist desperately looking for successful examples. You may become the light bearer for someone and affirm that another path, another world is possible.

International scientists also bring diversity of backgrounds, outlooks, and experiences. We model diverse forms of leadership. While there is much to learn from US society, we need not lose our authentic selves. We can integrate into US society while maintaining our unique strengths and identities. While learning and adapting, let us not forget to celebrate what we bring to your environments. Celebrate the international community and proclaim freely that global citizens shape the world of science.

Also, I urge you to simply be, to embrace your whole selves, your humanity beyond professional excellence. We can get caught up in proving our "extraordinary" or "outstanding" abilities as demanded by the US immigration system. Therefore, it will take intention to rest, to simply be, to breathe without abandon, to take up space and stop asking for permission. Your agency also lies in joy and restoration. A healthy mind and body can better imagine possibilities and life's varied purpose.

Finally, collectivism is the international community's superpower. Let's harness the collective empowerment model that centers on a culture of care to shape a curiosity-driven, inclusive, and humane world—within and beyond science. Together, we can make creative journeys and a better world possible.

BIBLIOGRAPHY

Alberts, Bruce, Marc W. Kirschner, Shirley Tilghman, and Harold Varmus. 2014. "Rescuing US Biomedical Research from Its Systemic Flaws." *Proceedings of the National Academy of Sciences* 111 (16): 5773–77. https://doi.org/10.1073/pnas.1404402111.

Amano, Tatsuya, Valeria Ramírez-Castañeda, Violeta Berdejo-Espinola, Israel Borokini, Shawan Chowdhury, Marina Golivets, and Juan David González-Trujillo et al. 2023. "The Manifold Costs of Being a Non-Native English Speaker in Science." *PLOS Biology* 21 (7): e3002184. https://doi.org/10.1371/journal.pbio.3002184.

American Immigration Council. 2022. "Foreign-Born STEM Workers in the United States." Fact Sheet. https://www.americanimmigrationcouncil.org/research/foreign-born-stem-workers-united-states.

Association of American Universities (AAU). 2019. "PhD Education Initiative." https://www.aau.edu/education-community-impact/graduate-education/phd-education-initiative.

Attig, Derek. 2017. "Using Job Ads for Career Exploration." *Inside Higher Ed.* https://www.insidehighered.com/advice/2017/04/03/examining-various-jobs-ads-can-help-you-explore-your-career-essay.

———. 2019. "The Strategic Way to Be Yourself in a Job Search." *Inside Higher Ed.* https://www.insidehighered.com/advice/2019/08/19/how-strike-balance-between-being-real-and-being-professional-job-interview-opinion.

Badlam, Justin, Stephen Clark, Suhrid Gajendragadk, Adi Kumar, Sara O'Rourke, and Dale Swartz. 2024. "The CHIPS and Science Act: What Is It and What Is in It?" McKinsey. https://www.mckinsey.com/industries /public-sector/our-insights/the-chips-and-science-act-heres-whats-in-it.

Bailey, Chris. 2017. "5 Research-Based Strategies for Overcoming Procrastination." *Harvard Business Review.* https://hbr.org/2017/10/5-research-based-strategies-for-overcoming-procrastination.

Bernstein, Shai, Rebecca Diamond, Abhisit Jiranaphawiboon, Timothy McQuade, and Beatriz Pousada. 2022. "The Contribution of High-Skilled Immigrants to Innovation in the United States." Working Paper Series. National Bureau of Economic Research. https://doi.org/10.3386/w30797.

Blake, Jessica. 2025. "What to Know about Trump's Strategy Targeting Colleges' Grants and Contracts." *Inside Higher Ed.* https://www.insidehighered .com/news/government/politics-elections/2025/04/18/what-know-about-trumps-funding-threats-colleges

Bipartisan Policy Center. 2022. "Reforming Employment-Based Immigration."https://bipartisanpolicy.org/report/reforming-employment-based-immigration/.

Burnett, Bill, and Dave Evans. 2016. *Designing Your Life: How to Build a Well-Lived, Joyful Life.* New York: Knopf. https://www.penguinrandomhouse .com/books/249885/designing-your-life-by-bill-burnett-and-dave-evans/.

Butler, Natalie. 2023. "Unlocking Potential of High Skilled Immigrants to Support the U.S. Economy." Bipartisan Policy Center. https://bipartisanpolicy .org/blog/unlocking-potential-high-skilled-immigrants/.

Carlsson, Sigrid V., Sandro C. Esteves, Elizabeth Grobet-Jeandin, Maria Chiara Masone, Maria J. Ribal, and Yao Zhu. 2024. "Being a Non-Native English Speaker in Science and Medicine." *Nature Reviews Urology* 21 (3): 127–32. https://doi.org/10.1038/s41585-023-00839-7.

Cassuto, Leonard, and Robert Weisbuch. 2021. *The New PhD.* Baltimore: Johns Hopkins University Press. https://doi.org/10.1353/book.81097.

Cépeda, Paola, and Natalie Chernets. 2022. "Setting up International Ph.D.s and Postdocs for Success." *Inside Higher Ed.* https://www.insidehighered .com/advice/2022/10/17/helping-international-trainees-prepare-jobs-us-opinion.

Chugh, Mayank, and Tiffany Joseph. 2024. "Citizenship Privilege Harms Science." *Nature* 628 (8008): 499–501. https://doi.org/10.1038/d41586-024-01080-x.

Collins, Bryan. 2021. "The Pomodoro Technique Explained." *Forbes.* https://www.forbes.com/sites/bryancollinseurope/2020/03/03/the-pomodoro-technique/?sh=2137ead83985.

Conrad, Jacinta. 2022. "Chips and Science Bill Isn't Enough. America Needs to Retain Its International Students." *APS News.* https://www.aps.org/apsnews/2022/08/chips-and-science-bill.

Cramer, Katryn D., and Hank Wasiak. 2006. *Change the Way You See Everything through Asset-Based Thinking.* Philadelphia: Running Press. https://www.hachettebookgroup.com/titles/hank-wasiak/change-the-way-you-see-everything-through-asset-based-thinking/9780762442119/?lens=running-press.

Crenshaw, Kimberle. 1989. "Demarginalizing the Intersection of Race and Sex: A Black Feminist Critique of Antidiscrimination Doctrine, Feminist Theory and Antiracist Politics." University of Chicago Legal Forum. https://chicagounbound.uchicago.edu/uclf/vol1989/iss1/8/.

Desikan, Anita, Jacob F. Carter, Abron Franklin, Raechel McKinley, Jennifer Orme-Zavaleta, and Andrew A. Rosenberg. 2023. "Diversifying the Federal STEM Workforce." Union of Concerned Scientists. https://doi.org/10.47923/2023.15242.

Drucker, Peter F. 2006. *The Practice of Management.* Reissue ed. New York: Harper Business.

Dweck, Carol S. 2006. *Mindset.* London: Penguin Random House.

Fagbule, O. F. 2018. "Use of Social Media to Enhance the Impact of Published Papers." *Annals of Ibadan Postgraduate Medicine* 16 (1): 1–2. https://www.ncbi.nlm.nih.gov/pmc/articles/PMC6143884/.

Fischer, Karen. 2023. "International Students Surge Back to U.S. Campuses." *Chronicle of Higher Education.* https://www.chronicle.com/article/international-students-surge-back-to-u-s-campuses.

Fisher, Roger, and William L. Ury. 1981. *Getting to Yes: Negotiating Agreements without Giving In.* London: Penguin Random House.

Fleming, Nic. 2022. "Underpaid and Overworked: Researchers Abroad Fall Prey to Bullying." *Nature* 608 (7922): 437–39. https://doi.org/10.1038/d41586-022-02142-8.

Foster, Bill. 2023. "Foster Leads Bipartisan Effort to Keep STEM Graduates in America." Press Release. http://foster.house.gov/media/press-releases/foster-leads-bipartisan-effort-keep-stem-graduates-america.

Froelich, Andrew. 2020. "What's the Difference between Social Media and Social Networking?" TechTarget. Search Unified Communications. https://www.techtarget.com/searchunifiedcommunications/answer/Whats-the-difference-between-social-media-and-social-networking.

Future of Research. 2017. "FoR Statement on President Trump's Executive Order on Immigration." https://www.futureofresearch.org/2020/07/20/for-statement-on-president-trumps-executive-order-on-immigration/.

FWD.Us. 2024. "Children of Immigrants in Green Card Backlogs Face Uncertain Futures in the U.S." Blog. https://www.fwd.us/news/children-of-immigrants-in-green-card-backlogs-face-uncertain-futures-in-the-u-s/.

Garriott, Omar, Jeremy Schifeling, and Lindsey Pollak. 2022. *Linked: Conquer LinkedIn. Get Your Dream Job. Own Your Future.* New York: Workman Publishing Company.

Goodman, Michael. 2016. "Systems Thinking: What, Why, When, Where, and How?" Systems Thinker. https://thesystemsthinker.com/systems-thinking-what-why-when-where-and-how/.

Han, Xueying, Galen Stocking, Matthew A. Gebbie, and Richard P. Appelbaum. 2015. "Will They Stay or Will They Go? International Graduate Students and Their Decisions to Stay or Leave the U.S. upon Graduation." *PLOS ONE* 10 (3): e0118183. https://doi.org/10.1371/journal.pone.0118183.

Inside Higher Ed. 2025. "International Student Visa Revoked." https://www.insidehighered.com/news/global/international-students-us/2025/04/07/where-students-have-had-their-visas-revoked.

Jorgensen, Christian, Miro Alexander Astore, Fátima Sanch.eznieto, Sridhar Vedachalam, Amreen Mughal, Maiko Le Lay, Adriana Bankston, and Harinder Singh. 2022. "Tracking Policy Implications and Impacts of the COVID-19 Pandemic and Related Executive Actions on a Sampling of Foreign-Born Early Career Researchers in the U.S." OSF. https://doi.org/10.31219/osf.io/n57vc.

Kahneman, Daniel. 2013. *Thinking, Fast and Slow.* New York: Farrar, Straus and Giroux.

Katz, Nora, and Hadeel M. Abouhasira. 2025. "Positive Changes for Business Immigration: The H-1B Modernization Rule." Holland & Knight (blog). https://www.hklaw.com/en/insights/publications/2025/01/positive-changes-for-business-immigration-the-h-1b-modernization-rule#:~:text=The%20Rule%20provides%20increased%20flexibility,met%20and%202)%20the%20individual.

Kaushik, Divyansh. 2023. "Understanding the Visa Policies under the AI Executive Order." Federation of American Scientists (blog). https://fas.org/publication/unlocking-american-competitiveness-ai-eo/.

Kuper, Julie L., and Jillian J. Turanovic. 2021. "Undoing Resilience: Immigrant Status and Poor Health Following Incarceration." *Health & Justice* 9 (1): 5. https://doi.org/10.1186/s40352-021-00129-7.

Langin, Katie. 2024. "Fewer U.S. Scientists Are Pursuing Postdoc Positions, New Data Show." *Science.* https://www.science.org/content/article/fewer-u-s-scientists-are-pursuing-postdoc-positions-new-data-show.

Majumdar, Sonali. 2021. "The Need for Self-Knowledge in a Changing World of Work." *Inside Higher Ed.* https://www.insidehighered.com/opinion/career-advice/carpe-careers/2023/02/26/lead-curiosity-and-creativity-your-career.

———. 2023. "Lead with Curiosity and Creativity in Your Career." *Inside Higher Ed.* https://www.insidehighered.com/opinion/career-advice/carpe-careers/2023/02/26/lead-curiosity-and-creativity-your-career.

Majumdar, Sonali, and Jenny Schneider. 2021. "Choice vs. Circumstance." *Inside Higher Ed.* https://www.insidehighered.com/advice/2021/01/18/three-distinct-strengths-contribute-professional-success-international-researchers.

Meadows, Donella H. 2008. *Thinking in Systems: A Primer.* White River Junction, VT: Chelsea Green Publishing.

Mentor Doctor (pseudonym). 2002. "Advisor versus Mentor." *Science.* https://doi: 10.1126/article.65490.

Milliken, Lindsay, and Simonai Santiago. 2021. "High-Skilled Immigration Policy: FAS's Current and Future Contributions." Federation of American Scientists. https://fas.org/publication/high-skilled-immigration-policy-fass-current-and-future-contributions/.

Mullaney, Thomas S., and Christopher Rea. 2022. *Where Research Begins: Choosing a Research Project That Matters to You (and the World).* Chicago Guides to Writing, Editing, and Publishing. Chicago: University of Chicago Press. https://press.uchicago.edu/ucp/books/book/chicago/W/bo131341275.html.

National Academies of Sciences, Engineering, and Medicine (NASEM). 2014. *The Postdoctoral Experience Revisited.* Washington, DC: National Academies Press. https://doi.org/10.17226/18982.

———. 2018. *Graduate STEM Education for the 21st Century.* Washington, DC: National Academies Press. https://doi.org/10.17226/25038.

———. 2019. *The Science of Effective Mentorship in STEM.* Washington, DC: National Academies Press. https://doi.org/10.17226/25568.

———. 2024. *International Talent Programs in the Changing Global Environment.* Washington, DC: National Academies Press. https://doi.org/10.17226/27787.

National Center for Faculty Diversity and Development (NCFDD). n.d. "The Mentoring Map." https://www.ncfdd.org/ncfddmentormap.

National Center for Science and Engineering Statistics (NCSES). 2023. "Doctorate Recipients from U.S. Universities: 2022." NSF 24–300. National Science Foundation. https://ncses.nsf.gov/pubs/nsf24300.

———. 2025. "Survey of Graduate Students and Postdoctorates in Science and Engineering: Fall 2023." NSF 25-317. https://ncses.nsf.gov/surveys/graduate-students-postdoctorates-s-e/2023.

National Institutes of Health (NIH). 2024. "NIH Advisory Committee to the Director, Working Group on Re-Envisioning NIH-Supported Postdoctoral Training: Report to the NIH Advisory Committee to the Director." https://acd.od.nih.gov/working-groups/postdocs.html.

National Postdoctoral Association (NPA) International Taskforce. 2023. "Onboarding International Postdocs Guide." https://www.nationalpostdoc.org/page/OnboardingInternationalPostdoc.

National Science Board, National Science Foundation. 2019. "Science and Engineering Indicators 2020: Science and Engineering Labor Force." NSB-2019-8. https://ncses.nsf.gov/pubs/nsb20198/.

———. 2022. "International STEM Talent Is Crucial for a Robust U.S. Economy." https://www.nsf.gov/nsb/sei/one-pagers/NSB-International-STEM-Talent-2022.pdf.

Nature. 2024. "Lands of Promise." Nature News & Comment. https://www.nature.com/news/7.6955?article=1.11602.

Nguyen, Thi A. 2018. "InterSECT Job Simulations." Science Sims. https://intersectjobsims.com/library/.

Okahana, H. 2019 "Closing Gaps in Our Knowledge of PhD Career Pathways: Job Changes of PhD Graduates after Earning Their Degree." CGS Research in Brief. Council of Graduate Schools. https://cgsnet.org/wp-content/uploads/2022/02/137_19-CGS_CareerPathways_July2019_WebFinal4.pdf.

Passport Index. 2024. "Global Passport Power Rank 2024: Global Mobility Intelligence." https://www.passportindex.org/byRank.php.

Ranieri, Anna. 2015. "Turning Your Complex Career Path into a Coherent Story." *Harvard Business Review*. https://hbr.org/2015/08/turning-your-complex-career-path-into-a-coherent-story.

Rathi, Anusha, and Christina Lu. 2023. "Chinese Scientists Are Leaving the United States." *Foreign Policy* (blog). https://foreignpolicy.com/2023/07/13/chinese-scientists-united-states-research-tech-academia-china-initiative/.

Roach, Michael, and Henry Sauermann. 2017. "The Declining Interest in an Academic Career." *PLOS ONE* 12 (9): e0184130. https://doi.org/10.1371/journal.pone.0184130.

Scellato, Giuseppe, Chiara Franzoni, and Paula Stephan. 2012. "Mobile Scientists and International Networks." National Bureau of Economic Research. https://doi.org/10.3386/w18613.

Sinche, Melanie. 2016. *Next Gen PhD*. Cambridge, MA: Harvard University Press. https://www.hup.harvard.edu/books/9780674986794.

Thompson, Glenn. 2012. "U.S. House Passes Thompson-Sponsored STEM Jobs Act." Press Release. http://thompson.house.gov/media-center/press-releases/us-house-passes-thompson-sponsored-stem-jobs-act.

United Nations Conference on Trade and Development (UNCTAD). 2023. Classifications—UNCTAD Handbook of Statistics. https://unctad.org/.

US Citizenship and Immigration Services (USCIS). 2022. "Employment-Based Immigration: First Preference EB-1." https://www.uscis.gov/working-in-the-united-states/permanent-workers/employment-based-immigration-first-preference-eb-1.

———. 2023a. "Chapter 4-O-1 Beneficiaries." https://www.uscis.gov/policy-manual/volume-2-part-m-chapter-4.

———. 2023b. "Updates. Child Status Protection Act (CSPA) Age Calculation for Certain Adjustment of Status Applicants." https://www.uscis.gov/newsroom/alerts/uscis-updates-child-status-protection-act-cspa-age-calculation-for-certain-adjustment-of-status.

———. n.d.a. "Eligible CIP Codes for the STEM OPT Extension | Study in the States." https://studyinthestates.dhs.gov/stem-opt-hub/additional-resources/eligible-cip-codes-for-the-stem-opt-extension.

———. n.d.b. "Exchange Visitors." https://www.uscis.gov/working-in-the-united-states/students-and-exchange-visitors/exchange-visitors.

———. n.d.c. "F-1 Curricular Practical Training (CPT) | Study in the States." https://studyinthestates.dhs.gov/sevis-help-hub/student-records/fm-student-employment/f-1-curricular-practical-training-cpt.

———. n.d.d. "Green Card for Employment-Based Immigrants." https://www.uscis.gov/green-card/green-card-eligibility/green-card-for-employment-based-immigrants.

———. n.d.e. "H-1B Specialty Occupations." https://www.uscis.gov/working-in-the-united-states/h-1b-specialty-occupations.

———. n.d.f. "O-1 Visa: Individuals with Extraordinary Ability or Achievement." https://www.uscis.gov/working-in-the-united-states/temporary-workers/o-1-visa-individuals-with-extraordinary-ability-or-achievement.

————. n.d.g. "Optional Practical Training (OPT) for F-1 Students." https://www.uscis.gov/working-in-the-united-states/students-and-exchange-visitors/optional-practical-training-opt-for-f-1-students.

US Department of Homeland Security (DHS). 2024. "Modernizing H-1B Requirements, Providing Flexibility in F-1 Program and Program Improvements Affecting Other Nonimmigrant Workers" *Federal Register.* https://www.federalregister.gov/documents/2024/12/18/2024-29354/modernizing-h-1b-requirements-providing-flexibility-in-the-f-1-program-and-program-improvements.

US Department of State (DOS). n.d.a. "Adjustments and Extensions. BridgeUSA." https://j1visa.state.gov/participants/current/adjustments-and-extensions/.

————. n.d.b. "New Initiatives Further Opportunity for International STEM Students, Scholars and Researchers, U.S. Entities." (Blog.) https://www.state.gov/new-initiatives-further-opportunity-for-international-stem-students-scholars-and-researchers-u-s-entities/.

————. n.d.c. "Opportunity for Academic Training Extensions for J-1 College and University Students in STEM Fields. BridgeUSA." https://j1visa.state.gov/opportunity-for-academic-training-extensions-for-j-1-college-and-university-students-in-stem-fields/.

————. n.d.d. "Waiver of the Exchange Visitor Two-Year Home-Country Physical Presence Requirement." https://travel.state.gov/content/travel/en/us-visas/study/exchange/waiver-of-the-exchange-visitor.html.

Vaccario, Giacomo, Luca Verginer, and Frank Schweitzer. 2021. "Reproducing Scientists' Mobility: A Data-Driven Model." *Scientific Reports* 11 (1): 10733. https://doi.org/10.1038/s41598-021-90281-9.

West, Darrell M. 2023. "Improving Workforce Development and STEM Education to Preserve America's Innovation Edge." Brookings Institution. https://www.brookings.edu/articles/improving-workforce-development-and-stem-education-to-preserve-americas-innovation-edge/.

White House. 2020. "Proclamation Suspending Entry of Aliens Who Present a Risk to the U.S. Labor Market Following the Coronavirus Outbreak." https://trumpwhitehouse.archives.gov/presidential-actions/proclamation-suspending-entry-aliens-present-risk-u-s-labor-market-following-coronavirus-outbreak/.

Willis, Deborah S. 2020. "Job Seeker, Change Agent." *Inside Higher Ed.* https://www.insidehighered.com/advice/2020/12/14/how-be-change-agent-diversity-equity-and-inclusion-your-career-and-job-search.

Witze, Alexandra. 2025. "75% of US Scientists Who Answered *Nature* Poll Consider Leaving." *Nature*. https://doi.org/10.1038/d41586-025-00938-y.

Xie, Yu, Xihong Lin, Ju Li, Qian He, and Junming Huang. 2023. "Caught in the Crossfire: Fears of Chinese-American Scientists." *Proceedings of the National Academy of Sciences* 120 (27): e2216248120. https://doi.org/10.1073/pnas.2216248120.

Zimmermann, Alessandra. 2023. "U.S. R&D and Innovation in a Global Context: The 2023 Data Update." American Association for the Advancement of Science. https://www.aaas.org/news/us-rd-and-innovation-global-context-2023-data-update.

INDEX

AAAS (American Association for the Advancement of Science), xiii, 28, 85, 148, 196, 207, 208
abuse in workplace, 4, 162–164, 203
academic administration field, xiv, 5, 45, 46, 48, 60
ACS (American Chemical Society), 28, 85, 196
Adjustment of Status (I-485) form, 106–107, 107
advisers: assistance with language, 60; building a network of, 70; collaborative grants with, 48; as distinct from mentors and sponsors, 70–71; exploring careers with, 85–86; IDPs and, 84; immigration portfolios and, 112, 114, 158; job search applications and, 118; mental well-being and, 158, 159, 162, 164; mentoring suggestions for, 193, 196–197, 199, 201–203, 205, 206; OPT and, 101; reliance for funding on, 12; seeking help from, 38; support from, 29, 182; systemic reforms and, 210, 212
Advisory Committee to the Director (ACD) (NIH), 216

advocacy: for evidence-based policies, 31; immigration system and, 107, 110; for international scientists, xiii, xv, 46, 158, 167; marginalization and, 168–169; mentoring and, 191, 201, 207; organizations for, 215–219; self-advocacy, 26, 51, 86, 137–138, 139, 153, 157, 190; self-knowledge and, 26, 85–86; social media and, 69; sponsors and, 71; for systemic change, 182, 183, 188, 206–207, 209
Alberts, Bruce, 208
Amano, Tatsuya, 57
American Association for the Advancement of Science (AAAS), xiii, 28, 85, 148, 196, 207, 208
American Association of Universities (AAU), 207, 208, 214
American Chemical Society (ACS), 28, 85, 196
American Society for Microbiology, 43, 142
Applicant Tracking System (ATS), 119, 121
Arab-Muslim internationals, 173. *See also* Middle Eastern internationals

tailored support for international scientists and, 3–8; visa sponsorship navigation, 5–6, 7; volunteer experiences, 4

LMIC (lower-middle income) nations, 169, 174–175, 190

Lu, Christina, 141

Lundsteen, Natalie, 145

Magaldi, Thomas (Tom), 5

Majumdar, Sonali, 25, 26, 28, 30, 50, 56, 85

Malik, Harmit, 109, 158, 179, 188, 191–193, 203, 204, 205–206

management by objectives (MBO), 88–89

managers: as creative partners, 200; EB-1C category and, 105; immigration requirements for trainees, 197; L-1 visas and, 151; lived experiences, 30, 45, 142, 143, 145; marginalization and, 168; mental well-being and, 203–204; mental well-being of internationals and, 158, 159, 167; mentoring suggestions, 112, 187–188; modeling good behavior, 206–207; suggestions for effective onboarding, 194; support for international employees and trainees by, 182. *See also* hiring managers; mentoring suggestions

Marballi, Ketan, 142–144, 147, 151–152

marginalization factors: about, 168–169; country of origin, 169–170; English-language fluency, 176–177; ethnicity and nationality, 172–173; gender and sexual orientation, 179–180; hidden curriculums, 178–179; immigration status, 170–172; lived experiences, 182–184; monolithic community and, 168; prestige of graduate and postdoc institutions, 177–178; recommendations and actions, 180–184; socioeconomic status, 173–176

Mastodon, 72

MBO (management by objectives), 88–89

McDowell, Gary, 161, 174, 182, 216

MedNess, 80

Memorial Sloan Kettering Cancer Center (MSKCC), 44

Memorial Sloan Kettering Cancer Center's Postdoc Association (MSK-PDA), 4

mental well-being: about, 157–158; global crises, 165–167; hustle for existence as unyielding, 160–161; identity struggles, 6–7, 165–167; lived experiences, 8, 160–161, 164; mentoring to support, 203–205; sacrifices of immigrant life, xv, 7, 8, 40, 158–160; workplace abuse, 4, 162–164

mentoring suggestions: about, 187–188; as community based; fostering growth mindset and creativity, 199–203; hidden curriculum, 197–199; hiring/recruiting international scientists, 188–190; key highlights, 207; lived experiences, 45, 201–203, 204, 205–206; modeling good behavior, 205–207; onboarding international scientists, 190–197; supporting diverse strengths, 203–205; supporting well-being, 203–205. *See also* community building

mentors: as distinct from advisers and sponsors, 70–71; lived experiences, 4, 5. *See also* professional mentorship

Middle Eastern internationals: global mobility, 8, 170; national security issues and, 165; xenophobia and, 173. *See also specific countries of origin*

Milliken, Lindsay, 216

MNCs (multinational companies), 152

model minority image, xv, 157, 173. *See also* humanity of international scientists

Founded in 1893,
UNIVERSITY OF CALIFORNIA PRESS
publishes bold, progressive books and journals
on topics in the arts, humanities, social sciences,
and natural sciences—with a focus on social
justice issues—that inspire thought and action
among readers worldwide.

The UC PRESS FOUNDATION
raises funds to uphold the press's vital role
as an independent, nonprofit publisher, and
receives philanthropic support from a wide
range of individuals and institutions—and from
committed readers like you. To learn more, visit
ucpress.edu/supportus.

www.ingramcontent.com/pod-product-compliance
Lightning Source LLC
Chambersburg PA
CBHW021920190326
41519CB00009B/859